Understanding Climate Change Adaptation

Praise for the book...

'This important book demonstrates that people are in the frontline of climate change around the world but are not simply passive victims – community-based adaptation is where the action is. It shows how engaged scholarship can both describe and harness the power of communities to face the challenge of climate change for generations to come. The work is a thoughtful and brilliant application of theory and practice in this crucial development arena.'
Neil Adger, *Professor, School of Environmental Sciences, University of East Anglia and Tyndall Centre for Climate Change Research*

'This book is a very timely attempt to analyse what adaptation actually looks like in practice and to set it in the context of climate change analysis and community dynamics.'
Richard Ewbank, *Climate Change Programme Coordinator, Christian Aid, UK*

'This book provides some excellent early learning from a number of case examples from different developing countries and will help those trying to understand and carry out community-based adaptation to climate change.'
Saleemul Huq, *Senior Fellow, Climate Change, International Institute for Environment and Development*

Understanding Climate Change Adaptation
Lessons from community-based approaches

Jonathan Ensor and Rachel Berger

Practical Action Publishing Ltd
Schumacher Centre for Technology and Development
Bourton on Dunsmore, Rugby,
Warwickshire CV23 9QZ, UK

www.practicalactionpublishing.org

© Practical Action Publishing, 2009

ISBN 978 1 85339 683 0

All rights reserved. No part of this publication may be reprinted or reproduced or utilized in any form or by any electronic, mechanical, or other means, now known or hereafter invented, including photocopying and recording, or in any information storage or retrieval system, without the written permission of the publishers.

A catalogue record for this book is available from the British Library.

The authors have asserted their rights under the Copyright Designs and Patents Act 1988 to be identified as authors of this work.

Since 1974, Practical Action Publishing (formerly Intermediate Technology Publications and ITDG Publishing) has published and disseminated books and information in support of international development work throughout the world. Practical Action Publishing Ltd (Company Reg. No. 1159018) is the wholly owned publishing company of Practical Action Ltd. Practical Action Publishing trades only in support of its parent charity objectives and any profits are covenanted back to Practical Action (Charity Reg. No. 247257, Group VAT Registration No. 880 9924 76).

Indexed by Indexing Specialists (UK) Ltd
Typeset by SJI Services
Printed by CLE Print Ltd
100% recycled FSC approved paper

Contents

Preface		vii
Acknowledgements		ix
1	Introduction: understanding community-based adaptation	1
	Abstract	1
	The development challenge	1
	The international context	4
	Climate predictions: understanding uncertainty	6
	Vulnerability and adaptation	13
	Adaptive capacity and resilience	17
	Knowledge, vulnerability and adaptation: a coherent approach	26
	Culture, communities and adaptation practice	33
	Structure of the book	36
2	Erosion and flooding in northern Bangladesh	39
	Abstract	39
	Introduction	39
	Community-based adaptation in the northern *charlands*	41
	Lessons and challenges	49
	Conclusion	52
3	Changing seasons and flash flooding in the foothills of the Nepal Himalaya	55
	Abstract	55
	Introduction	55
	Community-based adaptation in the middle hill region of Nepal (Chitwan District)	58
	Lessons and challenges	66
	Conclusion	68
4	Desert and floodplain adaptation in Pakistan	71
	Abstract	71
	Introduction	71
	Community-based adaptation in rural Pakistan	73
	Lessons and challenges	82
	Conclusion	85

5	Increasing paddy salinity in coastal Sri Lanka	87
	Abstract	87
	Introduction	87
	Community-based adaptation in coastal communities	89
	Lessons and challenges	97
	Conclusion	98
6	Increasing drought in arid and semi-arid Kenya	101
	Abstract	101
	Introduction	101
	Community-based adaptation in arid and semi-arid areas	103
	Lessons and challenges	111
	Conclusion	113
7	Multiple pressures on pastoralism in semi-arid Niger	115
	Abstract	115
	Introduction	115
	Community-based adaptation among the Tamasheq	117
	Lessons and challenges	124
	Conclusion	129
8	Declining water resources in Sudan's Red Sea coastal belt	131
	Abstract	131
	Introduction	131
	Adapting to drought in Arba'at	134
	Lessons and challenges	139
	Conclusion	143
9	Extreme weather in the Peruvian high Andes	147
	Abstract	147
	Introduction	147
	Community-based adaptation in Andean Peru	149
	Lessons and challenges	159
	Conclusion	161
10	Conclusion: community-based adaptation in practice	163
	Abstract	163
	The elements of adaptation	163
	Scaling up community-based adaptation	172
References		179
Index		185

Preface

Practical Action, established as Intermediate Technology Development Group (ITDG) in 1965, has framed an approach to community-based development that places people at the centre, focusing on the application of technologies appropriate to their situation to lift them out of poverty in ways that are sustainable over time in relation to the resources to which they have access. The majority of the work, managed from seven country and regional offices, is with rural communities and a significant proportion is focused on livelihoods based on natural resources: farming, livestock keeping, fishing and, to a minor extent, forests. From the late 1990s, communities were informing project staff that they were experiencing much greater variability in the weather and the timing of seasons than in recent decades. Extreme weather events were more frequent and more severe, so that coping strategies developed over many generations were being undermined: no sooner was recovery under way than another disaster struck.

In 2003, Practical Action was invited to a meeting in London convened by Saleemul Huq, of the International Institute for Environment and Development, an international research organization working on environment and development, and Andrew Simms, of the New Economics Foundation, a think-tank looking at alternative models of development. The meeting brought together many of the larger environmental and development NGOs to discuss climate change, and raised the question: was it not time for the two groups to work together to raise the profile of this issue with politicians in the UK, and more widely? At that time, Tony Blair, as British Prime Minister, was due to hold the joint presidency of the G8 and EU in 2005. That first meeting achieved the twin outcomes of an agreement of all the organizations to join together to advocate for the global temperature rise to be stabilized at no more than 2°C above pre-industrial levels, and to produce a report for launching in late 2004, drawing on the experiences of all the organizations in working with communities who were claiming to be affected by climate change. The report would highlight the dangers and challenges of climate change, but also put forward ways to help communities to cope and become more resilient.

The report that emerged was given the title *Up in Smoke* and was followed by three further regional reports focusing on Africa, Latin America and the Caribbean, and Asia. Each was launched in relation to a politically relevant event (the year of the Africa Commission, the anniversary of Hurricane Katrina, the UNFCCC COP in Bali), helping to bring the issue of adaptation onto the political agenda and increasing understanding of what adaptation means to a wider development audience. For Practical Action, it was with the preparation of case studies for the report that it became clear how relevant current field programmes in many regions were to the climate change debate. This in turn was the spur to look for funding for a project that would examine more specifically what constitutes adaptation, as opposed to sustainable development or disaster preparedness and prevention work. In 2004 Practical Action received funding from the Allachy Trust, a private UK funding body, to run a programme of work over three years on community-based adaptation in five countries in south Asia. In 2007, in its new five year strategy, Practical Action made climate change one of its two cross-organizational goals, reflecting an institutional commitment to helping poor communities to address the challenge of adaptation. This book is one result of this commitment, drawing together and reflecting on the experiences of Practical Action and others in working on community-based adaptation.

Acknowledgements

This book could not have been produced without the time and energy of the eight dedicated professionals who contributed to the case study chapters. These chapters are the result of full responses provided to a series of questions that were distributed by the authors in the opening months of 2008. The detailed material and subsequent correspondence provided by Cynthia Awuor (African Centre for Technology Studies, Kenya), Dinanath Bhandari (Practical Action Nepal), Dr Balgis Osman-Elasha (Higher Council for Environment and Natural Resources, Sudan), K.M. Mizanur Rahman (formerly Practical Action Bangladesh), Abdul Shakoor Sindhu (Rural Development Policy Institute, Pakistan), Juan Torres (Soluciones Prácticas, Peru), Rohana Weragoda (formerly Practical Action Sri Lanka) and Jeff Woodke (JEMED, Niger) has made it possible to publish this work with a view of adaptation in practice in eight countries. Invaluable feedback was provided by reviewers within Practical Action and by an anonymous external reviewer. The framework for understanding adaptation benefited from extensive discussions with Richard Ewbank of Christian Aid, and Pieter Van Den Ende, Katherine Pasteur, Andrew Scott, Hilary Warburton and in particular Barnaby Peacock at Practical Action in the UK.

CHAPTER 1
Introduction: understanding community-based adaptation

Abstract

This book is an attempt to further understanding of the practice of community-based adaptation. As such, the main body comprises a series of case studies of adaptation projects from some of the poorest regions of the world, placed between an introduction and conclusion that seek to clarify the meaning of adaptation and draw lessons for practitioners and policy makers. More specifically, this introduction has three objectives. The first is to locate the focus of this book – community-based adaptation – in the broader context of ongoing development challenges and the international climate change negotiations. This is the purpose of the following two sections. The second objective is to develop a framework for understanding adaptation. This is the task of the middle five sections, in which attention is drawn to the importance of uncertainty in climate predictions, the role of vulnerability in adaptation, and the components of adaptive capacity and resilience. The importance of an appreciation of the role played by culture in adaptation interventions is also highlighted. The concepts presented in these sections are illustrated with examples from the case study chapters that follow and are drawn together to offer a coherent approach to understanding adaptation. Third and finally, the last section provides an overview of the chapters that make up the remainder of the book.

The development challenge

Climate change first came on to the international political agenda in 1992 at the UN Conference on Environment and Development in Rio de Janeiro. It was regarded as an environmental problem and, accordingly, when the United Nations Framework Convention on Climate Change was agreed in 1995 government delegates were appointed from within environment ministries or meteorology offices. While the environmental movement had been raising the issue since the Intergovernmental Panel on Climate Change (IPCC) first assessment in 1990, climate change did not feature as an issue in mainstream development circles until the early part of the 21st century. Although humanitarian agencies and development organizations with a strong disaster relief focus were finding that they were facing one crisis after another – severe droughts in the Horn of Africa during the 1980s, Hurricane Mitch in Central America in 1998 and catastrophic floods in Bangladesh the same year

– these were not seen as evidence of a trend linked to global warming until the publication of the IPCC's Third Assessment Report in 2001. This indicated that increasingly severe weather events were likely to be a feature of global warming, and that global increases in sea surface temperatures contributed to the drying of the Sahel. What was evident from these disasters was that in each case those worst hit were the poorest people in a country or community, a fact that was brought to the world's attention when Hurricane Katrina struck New Orleans in 2005. It is this disproportionate impact on poor people that makes climate change an issue for development. It also forces us to ask why poor people are more vulnerable than other sections of the community who experience the same climate.

The most basic reason is poverty. Poor people are vulnerable to climate change because they have few assets and little to fall back on after a shock event which calls for resources to address it. They have no savings to replace or repair damaged property, and often no access to credit. Their ability to adapt to changed circumstances and adopt different livelihood strategies is limited, because they have little access to new knowledge or opportunities for learning new skills, and no capital to cover the costs of moving or setting up a new way of life. They also tend to have little access to and influence over the institutions and policies that affect their access to resources. The second reason is that poor people in rural communities depend on natural resources for their livelihoods: most of their income or production is from farming, fishing and livestock rearing. These activities are in turn dependent on soil, water and plant life, all of which are being adversely affected by climate change. Finally, poor people often live in places that are remote from services and information, have poor productivity or are prone to disasters. Better off people can afford to avoid these marginal locations, which include steep eroding hillsides, flood plains and low lying coastal areas. Despite these challenges, people in low productivity or hazard-prone areas have adapted using their own capabilities, skills, knowledge and technologies. For example, pastoralists worldwide have developed livelihood strategies for survival in extremely harsh climates with few resources. They make use of diverse grazing patterns and browse for water resources with livestock breeds that are matched to the subtly different environments available to them. In times of severe environmental stress a web of social norms and kinship ties provide support for survival and recovery.

The small-scale farmers who constitute the majority of the rural poor are the same people who have developed complex production patterns based on seed conservation and breeding and a diversity of crops to act as insurance against unpredictable weather and harvests. Farmers worldwide over millennia have been the custodians of biodiversity, cross-breeding wild varieties of edible plants to develop desirable traits for particular situations – climate, soil, altitude, disease resistance and many other characteristics. Practices to manage soil, conserve water and to use wild plants for food, fuel, medicine and construction continue to provide resilience and self-sufficiency for communities with few cash resources and who are remote from markets and sources of external knowledge. Here, people rely predominantly on their local knowledge, unique to their culture or society. Also known as indigenous knowledge or traditional wisdom, these practices and customs have been passed from generation to generation, usually by word of mouth and cultural rituals. This knowledge forms the basis for agriculture, food preparation, health care, education, conservation and the wide range of other activities that have sustained rural societies and their environment in many parts of the world for centuries. A key element of this knowledge is a variety of risk management strategies to cope with uncertain situations, such as growing certain crops as a form of insurance. Drought resistant millet and sorghum, for example, are planted in some areas so that if the main maize crop fails the low yielding but resilient crops still provide some food. Diversifying the crops grown and having multiple sources of food including wild plants thus reduces risks of food insecurity. Maintaining strong social networks and exchange relations with others in the local community is also an insurance strategy based on reciprocity in times of need.

The new challenge facing poor people is that climate change brings a further threat to a natural resource base that has become severely diminished in its ability to support an ever-increasing human population. As the Millennium Ecosystem Assessment (2005) revealed, ecosystems are facing severe decline in species diversity and resilience to shocks, through a combination of population growth, groundwater extraction for intensive agriculture, deforestation for timber and agricultural expansion, and widespread intensification of agriculture that has led to soil erosion and loss of biodiversity. All plants have a range of conditions they can tolerate, and many flourish in quite a narrow range of rainfall and temperature conditions. As climate changes emerge critical points will be reached where, in a particular locality, certain species fail to reproduce and become scarcer. In the Sahel and the Horn of Africa, for example, some pasture species valuable for their nutritive qualities for livestock are disappearing as a result of recurrent severe droughts, being replaced by opportunistic but less nutritious species. Extreme weather – be it wind, heat or rain – erodes soils, reducing the productivity of harvest and grazing lands and forest resources. The livelihoods that rural communities depend on have often developed over generations and are based on an intimate knowledge of their local environments. These livelihoods are now becoming less secure because of these factors. The ability to find new livelihood strategies, alternative crops or breed livestock that are more productive on poorer quality grazing is in many cases beyond the experience contained within traditional knowledge and requires access to outside expertise and information on what is happening to the climate. Poor people do not have access to these resources.

The purpose of this book is to increase understanding of how rural communities can be supported as they struggle to face the new challenges that climate change brings to their environment.

The international context

The United Nations Framework Convention on Climate Change has as its goal the stabilization of greenhouse gases to prevent dangerous interference with the climate system. Recognizing the need for actions to address the impacts of climate change, Article 4 of the convention places an obligation on developed countries to assist developing countries with adaptation. The targets for global emissions reductions determine the climate context in which adaptation takes place: the most recent science confirms that, as the global temperature rises, increasingly serious impacts will be experienced in food production, water resources, human health and the rate of extinction of plant and animal species (IPCC, 2007: 16). It is at the international climate change negotiations where an agreement will determine whether climate change will be contained through concerted international action on emissions reduction. Mechanisms for raising funding for adaptation and the amounts of money to be allocated will also be decided. While there is a growing level of understanding of the imperative for adaptation as well as mitigation, as of late 2008 there is still no firm commitment on the part of wealthy countries to contribute significant sums of money to help those in developing countries to adapt. However, it is clear that a prerequisite for any agreement on large-scale adaptation funding will be an understanding of how the money will be spent within developing countries and methods for measuring its effectiveness at improving the adaptive capacity of the poorest and most vulnerable communities. There can be no measure of the effectiveness of programmes for adaptation without a common understanding of what adaptation to climate change encompasses.

At the time of writing, there is little more than a year before a new international agreement should be reached during negotiations in Copenhagen in 2009. There is a strong group of informed development, environmental and research organizations working together to further understanding on adaptation at multiple levels: to implement effective projects and programmes, to influence the negotiators, and to influence governments. While there is extensive experience of development and disaster preparedness in areas of harsh climate or subject to extreme weather events, the vast majority of this work has not taken explicit account of climate change – of the fact that future change is certain, but that the direction and pace of change is not known; that extreme events will become more frequent, but as yet their timing cannot be predicted with certainty. While not focused on the policy context, this book supports the search for an international agreement by contributing to the building of an understanding of adaptation, with a specific focus on rural communities.

Early research efforts suggest that costs of adaptation will be substantial, ranging from between US$9 bn and $41 bn per year (World Bank, 2006a; based on an assessment of the vulnerability to climate change of current investments) to at least $50 bn per year (Oxfam, 2007; based on scaling-up the costs of recent community-based adaptation projects, the urgent needs identified in National Adaptation Programmes of Action (NAPAs), and an estimate of hidden costs such as capacity building and measures against desertification). The most recent UN Human Development Report suggests an even higher figure, estimating that a total of $86 bn will be needed by 2015 (UNDP, 2007: 194, to climate-proof existing development investment, adapt poverty reduction to climate change, and to strengthen disaster recovery). In stark contrast to the scale of these figures, the total funds pledged under the UNFCCC are very small, with around $149 m received for adaptation and a further $65 m pledged (Müller, 2008: 7). Under the UNFCCC, money is collected by voluntary contributions to a series of funds established under the Marrakesh Accords at the Conference of the Parties to the UNFCCC in 2001. A study published in 2007 by the Stockholm Environment Institute concludes that these global funds are not only inadequate in terms of funding, but also in terms of efficiency, fairness and in responding to developing countries' needs. The application procedures are complex, and inadequate support has been given to developing countries trying to apply for funds (Möhner and Klein, 2007). Also under the UNFCCC is a new fund, whose governance was finally agreed at the 13th Conference of the Parties in Bali in 2007. The Adaptation Fund is the only fund designed to support 'concrete' adaptation projects, raised by a market mechanism independent of donor governments. The Fund currently relies on a 2 per cent levy applied to Clean Development Mechanism (CDM) trading, yet the World Bank anticipates that this will only generate between $100 m and $500 m by 2012 (World Bank, 2006b). As we go to press, the future architecture, sources and scale of adaptation financing remain unclear beyond an apparent acceptance of the scale of funds needed.

Funding to undertake pilot programmes on adaptation and for action research is still extremely limited, despite the urgency and scale of the need to develop workable solutions to help people adapt. For example, the World Bank has established the Climate Investment Funds, most of which is for funding clean technology in developing countries. Of the approximately $6 bn pledged, the UK's Department for International Development has allocated £800 m, part of which is intended for adaptation under a new Pilot Programme on Climate Resilience. It remains unclear how much of the funding that is currently available is additional to official development assistance (ODA), as required under the UNFCCC. Yet climate change adds to the cost of achieving the Millennium Development Goals and therefore funding for adaptation must be over and above that promised for development. As will be seen in the following chapters, in order to implement successful adaptation projects with extremely poor and marginalized communities, it will be necessary to first undertake activities that are part of basic development as laid out within the Millennium Development Goals.

Despite the current inadequacy of funding, this book is written on the assumption that the Copenhagen agreement will see the commitment of large sums of money for adaptation to enable the scaling up of the kind of community-based interventions described in the following chapters.

Much more work is needed to assess the likely costs of adaptation under different levels of greenhouse gas concentration: the costs of ensuring that infrastructure and government services are resilient to climate change; the costs of prevention of and recovery from climate-related disasters; and the costs for communities and households. Whatever agreement is achieved in 2009, work on these issues will be ongoing, yet important discussions on how to enable adaptation (issues of process, capacity building and technology) and how to ensure effective delivery (issues of governance, decentralization and empowerment) must also run in parallel. These political and institutional challenges are at the heart of community-based adaptation and are returned to in the conclusion to this book.

The remainder of this introduction develops a framework for understanding adaptation that depends on two key parameters: the clarity or uncertainty of existing climate predictions and the vulnerability of a community or household to a given climate change hazard. After examining uncertainty and vulnerability in more detail, the elements of adaptation are considered, with particular emphasis given to the role played by social networks in enabling knowledge sharing, access to resources and influence over policy. The principal adaptation activities are identified as vulnerability reduction, building adaptive capacity and strengthening resilience. These actions are presented in a setting that illustrates how clarity and vulnerability determine the appropriate mix of activities in a particular context. The aim is to provide a mechanism for understanding adaptation that ensures the actions to support communities facing climate change are selected only after their particular circumstances have been assessed. By drawing attention to the uncertainty that is inherent in predicting climate change impacts, adaptation is presented as a process through which communities become increasingly able to make informed choices about their lives and livelihoods.

Climate predictions: understanding uncertainty

Advances in climate science have enabled climate modelling to provide an unprecedented view of the future of the Earth system. The impact of greenhouse gas emissions is now beyond doubt, as is warming of the global climate throughout the coming century. However, the precise implications remain unclear: predictions of rainfall rates, the likely frequency of extreme weather events and regional changes in weather patterns cannot be made with certainty. This uncertainty is of central importance to adaptation. While mitigation activities are rightly driven by the need to avoid dangerous climate change, adaptation planning cannot proceed without first understanding what climate change means in a particular location. Indeed, it is all too easy to assume that adaptation can and should follow climate change predictions. This can be the case where the message from current observations and predictive models is clear and unambiguous, such as for glacial melting or sea-level rise (and even here, the rate of change is a subject of debate). But clear-cut cases are in the minority. In many contexts there is no agreement whether, for example, rainfall is likely to increase or reduce. What, then, should adaptation to climate change mean in these circumstances? This question is addressed in the following two sections of this chapter (see 'Vulnerability and adaptation' and 'Adaptive capacity and resilience', below). First, it is necessary to consider the limitations of climate predictions so that the different aspects of uncertainty can be absorbed into adaptation thinking.

The Intergovernmental Panel on Climate Change (IPCC) provides the most well known and authoritative assessments of the current scientific understanding of climate change. The body was established in 1988 with a mandate 'to assess on a comprehensive, objective, open and transparent basis the latest scientific, technical and socio-economic literature' relevant to climate change (IPCC, 1988). A creation of the World Meteorological Organization (WMO) and the United Nations Environment Programme (UNEP), it conducts no new research of its own, but instead employs the services of around 400 scientists to compile reports on the 'policy relevant'[1] aspects of climate science, impacts and adaptation, and mitigation. Each report examines data from previously published peer-reviewed literature (and selected non-peer-reviewed reports) and is itself subjected to two rounds of expert review and one of government challenge and approval prior to publication. This approach removes controversial or spurious data and establishes a high degree of confidence in the content of the IPCC's publications. However, when relying on the IPCC's conclusions it is important to note that the approach to knowledge gathering and sharing has the potential to be conservative. Consensus building and time constraints mean that some evidence is excluded. For example, the IPCC's fourth report only considers temperature projections that fall within a 90 per cent confidence interval, excludes dynamic melting of the Greenland and Antarctic ice sheets, and excludes non-linear events that might result in higher or more rapid temperature or sea-level rise (Lemons, 2007). These decisions prevent the IPCC from reporting speculative data and allow scientifically plausible statements of certainty to be made in one report – but they also mean that low probability, high-impact events are not drawn out for public consideration. The periodic release of the IPCC assessment reports is also an important limitation: the time taken by the IPCC review process means that the evidence relied on in the reports is restricted to that published well in advance of the IPCC release date, while the 5 to 6 year periodicity of reports means that the most recent IPCC assessment lags behind current scientific thinking.

Rather than undermining the IPCC conclusions or the importance of the process, these observations highlight the need to understand the limits that inevitably exist to even the most authoritative statements of knowledge. Indeed, a 2008 review of the climate science literature illustrates the constraints of the IPCC process (Hare, 2008: 5–6):

Literature published in the past two years has identified several specific cases of higher risk than that assessed in the IPCC's AR4 [Fourth Assessment Report] ... this literature is sufficiently important, credible and robust to justify presenting a view that adds to, and in some cases differs from, the IPCC assessment. The reader should be aware, also, that this paper presents the science of climate change from a risk perspective, in terms of which low-probability, high-consequence events merit the attention of policymakers at the highest level.

For adaptation, the issue of interest is our ability to use science to predict the future. This is an obvious area for uncertainty to arise, and it does so in many forms. Fundamentally, the relationship between human activity and climate change means that assumptions must be made about the pattern of future emissions in order to generate climate predictions. Climate change predictions are mainly dependent on the greenhouse gas composition of the atmosphere in the future (predominantly carbon dioxide, methane and nitrous oxide). To address this problem, projections are made for a range of reasonably foreseeable future emissions. For example, the IPCC's best estimate for the increase in global average temperature by the end of this century is 1.8°C assuming a low rate of emissions (referred to by the IPCC as the B1 scenario), while the highest foreseeable increases in greenhouse gases would yield a 4.0°C temperature rise (the A1F1 scenario) (Meehl et al., 2007: 749).

The mechanisms involved in producing predictions also introduce uncertainty. Climate predictions are significantly different from their more established cousin, weather forecasting. Climate is fundamentally different from weather in that climate refers to long-term (conventionally 20 to 30 year) average weather conditions. Weather, on the other hand, refers to short-term (hourly and daily) changes such as in temperature, rainfall and wind. Weather is hard to predict as its dynamics are chaotic: small changes in the current weather conditions can create large changes in the weather at a later time. Despite this, well-established scientific understanding and measurement infrastructure allows predictions up to about 15 days ahead. Seasonal forecasting has emerged more recently than weather forecasting, and is based on an improved understanding of slowly changing phenomena that have a significant impact on the weather, such as the El Niño Southern Oscillation (ENSO). Measuring these important but slowly changing phenomena allows seasonal trends to be predicted up to around two years in advance, although confidence is greater for shorter timescales of up to around three months. Seasonal forecasts are not weather forecasts, but are more similar to climate models in that they offer a view of weather statistics, but over shorter timescales. A typical seasonal forecast may predict daily rainfall for a particular three month period. An expression of confidence is normally included: for example, a forecast may predict with 90 per cent confidence that daily rainfall will be between 150 mm and 200 mm. The confidence captures and communicates the uncertainty in the forecast, and varies significantly with geographical location: the more the weather is dominated by El Niño, for example, the more accurate the seasonal forecast. Generally speaking, predictability reduces the further a location is from the equator and from the ocean, and temperature is usually easier to predict than precipitation (Harrison et al., 2007a: 10).

Beyond seasonal timescales, climate models are relied on to provide information on long-term trends. While they are able to establish with high confidence that global average temperatures will continue to increase (not least because of the levels of greenhouse gases currently in the atmosphere), more detailed changes, such as the impact of warming on wet and dry seasons, remain unclear. The IPCC's calibrated expressions of confidence draw attention to the inherent uncertainty in climate models, which by definition are only approximations of reality, offering an incomplete representation of the full complexity of the Earth's systems. Uncertainty – meaning that more than one plausible future can be asserted – is unavoidable. Even for a fixed rate of future emissions there is uncertainty as to the exact impact on temperature. While the highest emission scenario produces a most likely average temperature increase of 4°C by the end of the 21st century, it is also possible that the increase might be as high as 6.4°C or as low as 2.4°C (Meehl et al., 2007: 749). Currently, the impact of uncertainty can be seen most clearly in the failure of climate models to provide good agreement at the regional scale, and in particular on future levels of precipitation.

Table 1.1 illustrates the extent of the problems with predicting precipitation changes from climate models. The table summarizes the climate change predictions for East Africa for the period 2080 to 2099, published in the IPCC's Fourth Assessment Report. They were generated by running 21 different climate models for a fixed increase in greenhouse gases.[2] The table compares the model responses to the data for 1980 to 1999, reporting the predicted change in temperature and percentage change in precipitation. The 21 models generate a spread of predictions as a result of the differences in assumptions and computational methods used. This diversity is captured in the following ways:

- The table shows the minimum (min), maximum (max), median (50 per cent) and quartile (25 per cent and 75 per cent) values from the 21 models for both temperature and precipitation change. Half of the 21 model predictions lie between the 25 per cent and 75 per cent figures. For example, half of the distribution of winter (December, January, February) temperature increases lie in the region 2.6 to 3.4°C, while the median (or middle) value from the 21 model outputs was a rise of 3.1°C. It is noteworthy that, while for all seasons the median precipitation change prediction suggests an increase, one or more models predict a decrease (demonstrated by the negative minimum values for this response).
- The quantity 'Tyrs' gives an estimate of when a clearly discernible trend emerges in the data. It gives the time, in years, before the 20 year average change in temperature or precipitation predicted by the models is greater

than the annual or seasonal variability generated by the models. For the temperature responses, the increasing trend is visible for all seasons after 10 years: in other words, it can be said with 95 per cent confidence that the average temperature in the period 1989 to 2009 will be greater than the average between 1980 and 1999.

- The predictions for precipitation are much less clear cut than for temperature: even during the winter season it is 55 years before there is 95 per cent confidence that the average precipitation will have increased. For the spring season the figure is greater than 100 years, indicating that even the predicted change at the end of the century cannot be expressed with confidence. During June, July and August no figure is given as half of the models have failed to agree on whether there will be an increase or decrease in precipitation by the 2080 to 2099 period.

This example demonstrates that timescales of a generation or more can be required before a clear pattern emerges from a suite of precipitation models, reflecting the uncertainty associated with predictions at the regional scale. The shortcomings of regional climate projections are well recognized in the scientific community. The World Climate Research Programme's Modelling Panel has concluded that 'regional projections from the current generation of climate models [are] sufficiently uncertain to compromise [the] goal of providing society with reliable predictions of regional climate change' (World Climate Research Programme, 2008a). As a result, attention is now turning to the possibility of large increases in computing power, with a proposal for a new billion dollar facility now on the table (Heffernan, 2008: 268). Following a summit meeting of international climate scientists in May 2008, a statement called for the establishment of an international 'Climate Prediction Project', the goal of which is to improve climate information to underpin 'regional adaptation and decision making in the 21st century' (World Climate Research Programme, 2008b). The hope is that improved computing power combined with investment in national research centres and improvements in scientific understanding will allow more precise regional data to be developed. However, such an initiative will have to overcome fundamental concerns that

Table 1.1 Temperature and precipitation predictions for East Africa

East Africa	Temperature change °C						Precipitation change %					
Season 2080–2099	Min	25	50	75	Max	Tyrs	Min	25	50	75	Max	Tyrs
Dec–Feb	+2.0	+2.6	+3.1	+3.4	+4.2	10	−3	+6	+13	+16	+33	55
Mar–May	+1.7	+2.7	+3.2	+3.5	+4.5	10	−9	+2	+6	+9	+20	>100
Jun–Aug	+1.6	+2.7	+3.4	+3.6	+4.7	10	−18	−2	+4	+7	+16	
Sep–Nov	+1.9	+2.6	+3.1	+3.6	+4.3	10	−10	+3	+7	+13	+38	95
Annual	+1.8	+2.5	+3.2	+3.4	+4.3	10	−3	+2	+7	+11	+25	60

Source: adapted from Christensen et al. (2007: 854)

remain among climate scientists. Problems with current models include their inability to reproduce tropical rainfall patterns, simulate the Arctic cycle, or mimic Atlantic hurricanes or European droughts (Pearce, 2008b). Moreover, the confidence that exists in current predictions is predominantly due to agreement being found between the predictions of several different climate models – such as where there is close agreement between the 25 per cent and 75 per cent columns in Table 1.1. However, this notion of multimodel consensus is itself open to question, the more so following studies that reveal the possibility of 'systematic error common to all models' – in essence, biases within the entire suite of models that undermine the reliability of the consensus (Palmer, 2008; see also Pearce, 2008a). Improvements are therefore necessary in the science behind climate modelling: massive investment in computing power alone will not be enough.

Using climate and forecasting information

Inevitably, all models will harbour uncertainty. Desai et al. (2008: 52) go as far as to suggest that models can only be 'heuristic tools which help our understanding of what we observe, measure or estimate' and should not be treated as 'truth machines'. If this is the case, then relying on climate models to provide information of sufficient precision to drive adaptation may always be a mistake. However, for the present, the question does not arise as model predictions remain either too contradictory or too broad to provide sufficient detail for a purely impacts-driven approach to adaptation. Indeed, it should be noted that moving from climate model outputs to a prediction of the likely impacts of climate change requires a further layer of modelling and therefore uncertainty. Impact models are a separate area of study from climate science that rely on physical and socio-economic models to translate a climate future (changes in temperature, rainfall and length of growing season, for example) into human impacts (such as health implications, flooding and food supply).

The aspects of uncertainty presented here should not be confused with a lack of knowledge: the issue is to develop a clear understanding of what climate science is currently offering. Changes in the Earth's systems are beyond doubt and as time passes the emerging science continues to suggest that the changes may be more profound and with us sooner than first thought. Climate change is already being experienced in many parts of the world, not least through rising sea levels, and climate predictions are clear in anticipating change at an unprecedented pace and scale. Yet the uncertainties associated with predictions are multifaceted and complex. If adaptation is to respond to the challenges of climate change, both these aspects of climate knowledge – the process of change and the degree of uncertainty – must be understood. Some level of understanding of what climate change may bring in a particular setting and how uncertain those predictions are will be essential if informed decisions are to be taken.

It follows that short-term forecasting and climate predictions both have a role in adaptation. This is reinforced in the case study chapters that follow, where it is found that local knowledge, which traditionally underpins community responses to changes in the weather and seasons, needs increasing support in the face of unprecedented climate variability. The different challenges that climate change presents are best addressed through different approaches to forecasting or prediction. Extremely fast onset events, such as flash floods or cyclones, require weather forecasting (looking at the days or weeks ahead); fast onset events, such as drought, need seasonal forecasting (at a timescale of weeks and months); and slow onset events, such as changes to seasons or sea level, can to an extent be predicted through climate modelling (looking at multiple years into the future) (R. Ewbank, personal communication, 26 September 2008). Yet there are also important links across these timescales, as changes to the climate place short-term events in a new context. Forecasting underpins the early warning that is necessary to ride out extreme weather events, be it through temporary relocation, securing property or ensuring adequate food supply. Climate change model predictions, on the other hand, may indicate the future likelihood of, for example, high rainfall or temperatures. This allows current unexpected events to be understood as part of an emerging trend (rather than as an anomaly) and offers a new context for infrastructure investment or livelihood decision making. This holds equally for incremental changes (such as gradual changes in temperature, sea level or rainfall), which can go unnoticed, be perceived as anomalous, or be disguised in or misdiagnosed as part of cyclical climate variations. Such changes may be foreseeable through seasonal forecasting but must also be recognized as part of a new long-term trend if they are to be understood and planned for.

Thus it is through short-term and seasonal forecasting that adaptation actions are able to engage with the variable conditions that are superimposed on emerging climate trends. However, a risk remains that simple climate change messages can disguise the complexity that is inherent in climate modelling. This applies to uncertainty (where similar outcomes can be equally asserted) and the statistical nature of predictions (in that climate only offers a view of the average conditions over a long period). If the uncertainty in climate messages is overlooked or underplayed then actions may be driven towards maladaptations focused on predicted impacts that fail to materialize. Equally, however, complexity and uncertainty must not divert adaptation from addressing threats to livelihoods if such impacts are known and imminent. These challenges are reflected in the Kenya case study reported in Chapter 6. In a context dominated by reductions in annual rainfall, the need for information was met through efforts to translate seasonal forecasting into locally relevant agricultural information. While the need to inform the community about long-term climate change was recognized, specific predictions were at the time judged to be too uncertain to have a role in adaptation actions.

In short, relevant and appropriate dissemination of climate change-related information, encompassing weather forecasts, seasonal forecasts and climate change predictions, is an important – if not defining – feature of adaptation.

Vulnerability and adaptation

Vulnerability underpins adaptation: it is the perception of vulnerability to some aspect of climate change that motivates and defines the objectives of adaptation activities. However, this observation raises two questions. First, how is this perception arrived at? As indicated above, the generation and communication of climate change knowledge plays a significant role in how we understand the conclusions of climate science, and this topic will be considered further in the following section (see 'Adaptive capacity and resilience', below). The second question is more fundamental: what do we mean by vulnerability? This question is the subject of the following paragraphs, in which alternative perspectives on vulnerability are shown to differ in their treatment of the uncertainty in climate change predictions, and yield different approaches to adaptation.

Before moving to consider vulnerability in detail, it is first necessary to ask 'vulnerability to what?' An answer of 'climate change' is only a starting point, as climate change will bring very different impacts in different places. However, three qualitatively different phenomena can be identified, offering a categorization that can help with understanding the challenges of climate change (Brooks, 2003: 9):

- *Category 1*. Discrete recurrent hazards, as in the case of transient phenomena such as storms, droughts and extreme rainfall events.
- *Category 2*. Continuous hazards, for example increases in mean temperatures or decreases in mean rainfall occurring over many years or decades.
- *Category 3*. Discrete singular hazards, for example shifts in climatic regimes associated with changes in ocean circulation.

While climate modelling makes clear that the possibility of major environmental changes is real, category 3 hazards are currently poorly understood and their possible impacts for communities unknown. Examples of the first two hazard categories, however, are found in the case studies examined in this book. The challenge for adaptation is very different for each of these classes of hazard. Category 1 hazards, for example, may demand early warning or disaster risk reduction approaches, category 2 the adoption or evolution of new livelihood practices, while category 3 may demand the abandonment of existing lifestyles. In any particular case, however, it remains necessary to consider vulnerability to a particular hazard or category of hazards.

Climate change hazards yield two forms of impact: biophysical impacts, such as the physiological effects on crops, changes in disease vectors, or changes to land, soil and water quality and quantity; and livelihood (or socio-economic) impacts, including damage to infrastructure or changes in crop production or trade patterns (Orindi and Murray, 2005: 5–6; FAO, 2007: 2). Crucially, livelihood impacts arise as a result of the interaction of biophysical impacts with existing social systems. For example, mortality rates following the emergence of new diseases are dependent on factors such as access to health care and nutritional intake. Similarly, a change in land quality (a biophysical impact of climate change) interacts with a community or household's existing social and economic systems to be translated into livelihood impacts such as a fall in income from crop production, reduced food security or human migration. The case of Bangladesh, examined in Chapter 2, illustrates this interaction: persistent flooding and high river flow rates erode the fertile land that local people depend on for their livelihoods, while relocation to islands created by the change in river flow is restricted by local landowners. Clearly, it is the prospect of livelihood impacts that is central to motivating adaptation, yet differences in how livelihood impacts are understood generate two quite different versions of vulnerability.

The first approach to vulnerability arose from a desire to estimate the potential costs of climate change (Burton et al., 2002). In this view, vulnerability refers to the 'end-point' livelihood impacts of climate change after an adaptation option has been adopted (Kelly and Adger, 2000: 327). Rather than assessing context, end-point vulnerability seeks to measure the effectiveness of an adaptation option in reducing the damage brought about by a particular hazard. For example, the difference in yield between two proposed crop varieties after flooding quantifies relative vulnerability. This approach facilitates the assessment of competing adaptation interventions based on the outcome of an anticipated climate hazard: the introduction of the crop with the highest yield would be the chosen adaptation. By contrast, the second approach – the 'starting-point' interpretation – focuses on understanding the processes that pre-exist within a system (Kelly and Adger, 2000: 327; O'Brien et al., 2004: 2). It seeks to examine the characteristics of communities to identify those elements that make them susceptible to climate change: in this sense the approach considers vulnerability to exist independently of the external climate hazard. Thus rather than suggesting the introduction of the highest yielding crop, an assessment of starting-point vulnerability should address the breadth of issues that play a role in vulnerability to flooding (such as the lack of watershed management cooperation between upstream and downstream communities, illegal logging or degraded soils with poor moisture retention). Alternative crop varieties would also be subject to this analysis (for example, whether the seeds can be stored and reused on farm, and the cost and availability of the inputs necessary to obtain high yields). This explicit focus on the broad social and environmental context, rather than on a particular adaptation intervention, distinguishes starting-point from end-point vulnerability.

The starting point definition of vulnerability is assumed throughout this book. Understood in this way, vulnerability is determined by the environmental and human characteristics of the community, revealing the process through which climate change hazards generate livelihood impact. Human factors such as food entitlements and access to health care will play an important role in determining the outcome of several hazards, while others, such as the quality or location of housing, may only assume significance for a particular hazard (for example, flooding). Environmental variables, such as topography, biodiversity or groundwater reserves are key to mediating climate hazards, moderating changes in all aspects of weather including temperature, precipitation and extreme events. These variables are also influenced by human activities as communities strive to manage and exploit the environment (Brooks, 2003: 5). Methods for assessing starting point vulnerability are explored in several of the following chapters. For example, in Sri Lanka (Chapter 5) the project team engaged in several activities with the community including resource mapping, risk mapping, and field observations and transect walks. Each of these approaches yielded different information about context in which risks and threats to the local resource base, environment and livelihoods have emerged. The Pakistan case study (Chapter 4) notes the importance of gender analysis in the process of assessing vulnerability, as women's livelihoods often have different characteristics from those of men. Women also offer a different insight into the local threats and risks.

There are several reasons for adopting the starting point definition of vulnerability. Starting-point vulnerability recognizes that ongoing livelihood activities, coping mechanisms or broader issues (such as the emergence of conflict) can generate changes in the human and environmental factors that underpin vulnerability (Smit and Wandel, 2006: 287). This dynamic aspect of vulnerability – the possibility of change to the social, political, economic and environmental characteristics of a community – is only captured in the starting point description. Yet capturing changes to vulnerability over time is essential when considering climate hazards. Category 1 (discrete recurrent hazards) and category 2 (continuous hazards) both draw attention to climate change as an ongoing phenomenon (with impacts experienced either intermittently or progressively). A meaningful representation of vulnerability to climate change must therefore capture the incremental changes to vulnerability that result from multiple or progressive impacts over time. Moreover, there are inherent difficulties with end-point analysis that are not immediately obvious. End-point vulnerability is forward looking in the sense that it tries to anticipate the outcome of climate hazards. It therefore relies on social and economic models to translate *predicted* biophysical impacts into *predicted* livelihood impacts following some *anticipated* adaptation: at each stage uncertainty is introduced and compounded. Moreover, the reliance on social and economic models inevitably increases as the analysis moves further into the future and climate hazards extend beyond the realm of experience. As Burton et al. (2002: 8) state, '[a]nalysis of the choice of adaptation measures at some future time to an uncertain future climate in an unknown socio economic context is bound to be highly speculative'.

A further problem arises when the class of adaptation activities associated with end-point vulnerability are considered. These are referred to as first generation or standard approaches (Burton et al., 2002: 7) and aim to reduce end-point vulnerability by finding a particular adaptation solution to a particular climate change problem. Typically, first generation approaches have been associated with technological fixes, such as raised bridges, improved levees or new monoculture crop varieties. However, the 'speculative' nature of the analysis can lead to overwhelming uncertainty and ultimately to maladaptations if the reality of climate change turns out to be different from current expectations (O'Brien et al., 2004: 5). Second generation adaptation takes a very different approach, commencing from the premise that it is necessary to address the context in which hazards occur (Burton et al., 2002; O'Brien et al., 2004; Eriksen et al., 2007). Starting-point vulnerability is employed to examine the biophysical, social, economic, political and cultural factors that make up climate vulnerability. By shifting focus in this way, the function of adaptation becomes one of reducing the causes of vulnerability when addressing particular outcomes. The intention is to reduce current vulnerability to the climate change hazard (or hazards) without the risk of maladaptation: activities should simultaneously reduce the impacts of potential climate change and improve the well-being of households or communities in the short term – for example through addressing poor housing, degraded soils or the inequitable distribution of resources. These approaches to adaptation are in essence vulnerability reduction measures, and can be classified as no-regrets strategies – meeting climate change adaptation goals while fulfilling broader development ends even if climate change predictions do not play out. Many of the approaches to adaptation examined in the following chapters employ no-regrets strategies. For example, in Pakistan (Chapter 4), small-scale vegetable farming provided food security and increased income almost immediately, while also being much less water intensive than the local cash crops and thus well suited to the (anticipated) dryer, water scarce climate. Similar approaches were adopted in Nepal, Kenya and Sudan (Chapters 3, 6 and 8).

As the case studies in this book illustrate, the focus on the causes of vulnerability does not mean that second generation approaches cannot address livelihood impacts. Nor does it mean that starting point vulnerability can or should operate independently of climate change predictions. Analysis of vulnerability will need to be with reference to an anticipated hazard (for example, vulnerability to sea-level rise). Moreover, there will be circumstances when climate change predictions are clear or impacts are evident in the short term (such as glacial lake outburst or where sea-level rise leads to flooding). However, starting-point vulnerability should ensure that the chosen response (such as the introduction of saline tolerant varieties or dam building) recognizes the 'existing social, economic and political structures' and thus that adaptations 'may increase inequality in a community and exacerbate vulnerability for some' (O'Brien et al., 2004: 12).

Adaptive capacity and resilience

While starting-point vulnerability can underpin no-regrets approaches to adaptation, investment in these strategies inevitably requires a degree of confidence in climate change predictions. A point will arise when uncertainty about the future climate increases so much that vulnerability assessments start to lose value, eventually becoming meaningless as the real possibility of unforeseen hazards emerges. If increased rainfall is as likely to emerge as increased drought, how then should adaptation proceed? This section focuses on how this challenge can be met through building adaptive capacity, understood as the ability to change in response to climate changes, and resilience, understood as the ability to absorb or cope with the unexpected. However, adaptive capacity and resilience should not be seen as independent of vulnerability: increasing a household or community's resilience or ability to adapt should help reduce vulnerability to the broadest possible range of hazards. As discussed below, adaptive capacity is also related to resilience, as building adaptive capacity is one way to support the ability to cope and recover.

Adaptive capacity refers to the potential to adapt to the challenges posed by climate change, describing the ability to be actively involved in processes of change. It encompasses the ability of actors within a particular human and environmental system to respond to changes, shape changes and create changes in that system (Chapin et al., 2006: 16,641). The tools that make up adaptive capacity therefore include both tangible assets, such as financial and natural resources, and less tangible elements such as the skills and opportunities to make decisions and implement changes to livelihoods or lifestyle. Both the diversity and distribution of these components of adaptive capacity are important. For Chapin et al. (2006: 16,641) adaptive capacity: 'depends on the amount and diversity of social, economic, physical, and natural capital and on the social networks, institutions, and entitlements that govern how this capital is distributed and used'.

Similarly, Smit and Wandel's (2006: 286) review of adaptation literature suggests that the determinants of adaptive capacity include both assets, including financial, technological and information resources, and the context within which these assets are held, including infrastructure, institutional environment, political influence and kinship networks. Thus both Chapin and Smit and Wandel draw attention not just to the availability of assets but also to the prevailing social and political context through which distribution takes place: networks, institutions, entitlements and political influence. Smit and Wandel (2006: 289) note that this context operates at different scales: while some elements of adaptive capacity are local (such as networks of family relationships), it is also important to recognize that broader and sometimes global social, economic and political forces may have the most significant influence on local vulnerabilities, such as where international free trade agreements remove supportive subsidies or price guarantees for a particular local crop. It may not be sufficient to consider only micro-scale relationships if it is 'powerful political and economic vested interests that determine the nature of the adaptation context' (Brooks, 2003: 12).

Diversity supports adaptive capacity by providing communities with options at times of stress or external change. Diversity is an attribute that offers more than simple accumulation of assets: it recognizes that addressing an uncertain future requires access to a range of alternative strategies, some of which will prove viable. However, diversity is also a key pillar of resilience. Where adaptive capacity refers to the ability to influence and respond directly to processes of change, resilience is the ability to absorb shocks or ride out changes. For resilience, diversity of social, economic, physical and natural assets improves the prospects of a socio-ecological system persisting. For example, a farm system dependent on a single crop may have low resilience to disease or climate change compared with one predicated on agricultural biodiversity. Similarly, the Kenyan case study (Chapter 6) reports how a diversity of seeds are planted in response to uncertain seasonal forecast information, ensuring that some crops reach maturity even if the predicted weather patterns do not emerge. In Peru, a diversity of planting altitudes and terrains underpins the conservation of potato varieties from one year to the next, ensuring that crops survive in some locations even if they fail in others (Chapter 9). In the same way, resilience may also take the form of multiple (diverse) livelihoods. While diversity underpins resilience, it is also important to recognize that the scale or degree must be appropriate and that a point can be reached at which assets or skills are spread too thinly to be of benefit. In some circumstances sufficient accumulation of assets may also support resilience: reserves of financial capital, for example, can be enough to ensure a household can cope in many circumstances. Safety nets such as insurance, when available and affordable, can also form an important component of resilience and may play a role in backstopping specialization or compensating for a lack of diversity.

Resilience and adaptive capacity are closely related, not least because both reduce the impact of uncertainty. Fostering adaptive capacity is also a mechanism for building resilience: adaptive capacity expands the options and opportunities for coping with or avoiding the impacts of climate change and thus improves a community or household's prospects of survival. For example, given access to and the ability to employ climate-related information, a pastoralist community may proactively sell livestock prior to a drought, thus yielding the resources necessary to cope (as in the Niger case study in Chapter 7). In this example, resilience – the broad ability to cope and recover – was enhanced through adaptive capacity that enabled the community to engage with and respond to the prospect of drought.

Elements of adaptive capacity

Adaptation requires the accumulation of skills as well as a diversity or accumulation of assets. Principally, utilizing a diversity of assets to expand the range of available livelihood or coping strategies requires the ability to explore ways of employing those assets. Thus attributes of adaptive capacity also include the ability to experiment or innovate, and the capacity to learn (Peterson, 2000: 328; Chapin et al., 2006: 16,641). Indeed, it has been suggested that the most adaptive societies are those with actors who have the capacity to experiment, and institutions in place to support them (Patt, 2008). The involvement of NGOs and institutions in farmer-led research into rice varieties in Sri Lanka, explored in Chapter 5, demonstrates how local adaptive capacity – and in particular the confidence to experiment – can be fostered. A similar experience is reported in the Peru case study (Chapter 9). The provision of technical training is an important element in supporting experimentation and learning, and is present in many of the case studies. For example, training in raft construction in Bangladesh (Chapter 2) allowed local farmers to implement their own changes to the design of floating gardens, ensuring they were suitable for the local conditions. Local extension services can also be key, as demonstrated in the Pakistan case study (Chapter 4) in which the presence of NGO and extension support is identified as part of the difference between success in one site and failure in another. The readiness to experiment and learn is complex and influenced by human, cultural, financial and institutional factors. The ability to put a diversity of resources to productive use may be linked to educational background and prior experience. Cultural attitudes may overlap with attitudes to financial assets to assist or inhibit experimenting and risk taking. For example, evidence suggests that there is no correlation between farmers' wealth and their willingness to adopt new management practices (Phillips, 2003; A. Patt, personal communication, 6 June 2008). On the other hand, the take up of insurance – a method of reducing risk and thereby facilitating experimentation – has been found to be greater among wealthy households. Risk-averse households can in fact be less likely to take insurance, often because of their lack of experience with handling financial products (Gine et al., 2007: 2). Some social norms, if narrowly defined or deeply held, can stand in the way of experimentation. However, culture may equally support or inhibit experimentation: marginalized communities have exhibited both conservatism and experimentation as a strategy to deal with environmental change (Ensor and Berger, 2009; Patt, 2008: 63).

The process of learning and adopting new strategies can be closely linked to the presence of social networks (defined in more detail below). Gine et al.'s (2007: 19) study of insurance take up among rural households in Andhra Pradesh reveals membership of social networks to be important in determining whether a new insurance scheme is adopted, as networks provide opportunities for sharing information and advice. The importance of networks is explained in part by research which suggests that individuals respond differently to information that is gained through 'experience-based reasoning' or 'analytically-based reasoning'. Experience-based information processing is generally dominant in decision making, and relies on experience that is personally held or communicated from others. Analytical reasoning responds to authoritative, externally provided information and generally has a role restricted to moderating the experiential response (Balstad, 2008: 166). Social networks provide an opportunity for sharing experiences and therefore are well placed to be effective in promoting learning, influencing changes to behaviour and stimulating collaborative innovation processes (Cross and Parker, 2004: 9). In the same way, they can help the real and perceived risk of adopting changes to livelihoods to be reduced by observing and understanding the experiences of others. This is evident in many of the case studies examined in this book, including in Nepal (Chapter 3) where the practice of vegetable growing was introduced to demonstration sites in the first year of the project and spread within two years to almost all farmers in the area.

Importantly, adaptive capacity also requires the ability to access and process climate information. Climate change is an emerging phenomenon with the potential to transform environments and challenge traditional expectations of seasonal patterns and climate extremes. As adaptive capacity embodies a household or community's ability to engage with and make decisions about processes of change, some level of understanding of climate change predictions and the associated uncertainty is essential. Indeed, in most of the case study chapters that follow, raising awareness of climate change was a central adaptation activity. As noted, however, there is also the potential for simple climate change messages to disguise the complexity that is inherent in climate modelling. The illusion of certainty and simplicity is attractive. Social scientists and psychologists have found a tendency for people to reduce complexity by focusing on single problems and single solutions even when faced by multiple threats and where a diverse response may be more effective (Balstad, 2008: 167). This poses a problem for the communication of climate change information and raises the prospect of uncertainty being overlooked or underappreciated by either party when information is exchanged. However, an appreciation of complexity and uncertainty is essential if maladaptations or futile efforts at vulnerability reduction are to be avoided.

Some form of institutional support is necessary for information dissemination, as climate change science and predictions will be beyond the reach of most poor communities. It is the responsibility of national governments to assimilate and communicate short- and long-term weather and climate change information, and to identify and facilitate the filling of gaps in knowledge where they exist. Moreover, each of these responsibilities should be grounded in the livelihood context of those who are most vulnerable to climate change: information should be targeted at these groups in a form and with content that is appropriate to their needs. This responsibility can be framed either generally, in terms of social contract between government and citizens, or more specifically in terms of the rights violations that will result from the depletion of resources and loss of lives and livelihoods.

Social networks

Social networks can be of significance to adaptive capacity and resilience. Both demand a degree of collective action and depend on the particular web of relationships that determine power, resource and information distribution in any situation involving multiple stakeholders. The focus of social networks is on the nature of these relationships. This distinctive perspective has given rise to key concepts that can be used to analyse and understand the connections between different actors (Wasserman and Faust, 1994: 4):

- Actors and their actions are interdependent rather than independent, autonomous units.
- Relational ties (linkages) between actors are channels for transfer or flow of resources (either material or non-material).
- The network structural environment provides opportunities for, or constraints on, individual action.

In the network view it is the relationships that link actors that are central rather than the individuals themselves. Actors are interdependent, and it is through their relationships that they create opportunities for resource and information exchange, and form the social, economic and political structures that define how they as individuals or groups may act.

Network analysis reveals the nature and extent of the interconnections between actors within the network (Hawe et al., 2004). Social networks have much in common with social capital, a concept often referred to in development contexts. Adger (2003: 392), for example, draws attention to two forms of social relationships, defining social capital as either bonding or networking. Bonding capital refers to the often strong ties of friendship and kinship. Networking capital, however, focuses on relationships outside the immediate group between people with different backgrounds but shared interests, including vertical relationships such as those between communities and their governing elites. Networking capital is built on trust and reciprocity, such that positive behaviour is expected and replicated by members of the network, while destructive behaviour can lead to the breakdown of relationships. This form of social capital can therefore be weak or fragile, and is often situated in institutions that have formal rules of behaviour. Both these aspects of social capital make up part of the social network of every individual, household or community. Critically, however, the network view goes further, drawing attention to the power and interests that define the nature of the relationships between actors, and identifying the direct and indirect connections that channel flows of resources and information.

A community or household's social network plays a role in adaptation beyond the support of learning, experimenting and innovating described above. In the best case, a community's network will yield a productive, open and democratic relationship to the state, promoting both policy and social learning (Adger, 2003: 394) and illustrating the flow of non-material resources along network linkages. For adaptation, such a relationship between the state and community and/or civil society networks might facilitate the two-way flow of information: upwards from household and community to improve policy understanding of the local socio-economic and environmental context, knowledge and needs; and downwards in the delivery of relevant and current science. In Kenya (Chapter 6), the integration of meteorological office staff into the project team ensured that forecasting information was generated for the project area and converted into a format useful for local farmers (for example, appropriate crop varieties for the anticipated rainfall). In Sudan (Chapter 8) strong links between the community and state government enabled support for project activities through improved information, and are helping the community to have its views heard on the construction of a new dam. However, where a productive relationship between state and civil society is not present, networks generate adaptation benefits in their own right, providing opportunities for sharing knowledge and learning. Local social networks offer marginalized groups an opportunity to develop adaptive strategies in the face of emerging climate risks. Adger (2003: 399) describes the use of social networks connecting local users for adaptive natural resource management in the absence of state planning in Vietnam. This use of networks was also in evidence in Sri Lanka (Chapter 5), where local farmer groups became agents of change at the community level, sharing their experiences of rice variety research with farmers from neighbouring villages. Moreover, social networks are known to be relied on for coping during times of stress, most clearly through kinship ties (Pelling and High, 2005: 311) but also in reciprocal arrangements between members of a network (Adger, 2003: 399). Arrangements that allow help to be requested between families on condition that it will be reciprocated when requested are at the heart of the *ayllus* kinship social group system in Peru's harsh mountain environment (Chapter 9).

The process of building adaptive capacity may take the form of building on existing social networks. Depending on the context, this may be through working with local networks with a view to enhancing and sharing community-based adaptation activities. Alternatively, it may be through supporting civil society organizations to pressure different levels of government into participating in new (or existing) forums so that demands for climate information and adaptation resources can be articulated. Social network analysis supports these processes by systematically identifying the opportunities and constraints that exist in the links that make up the networked relationship between communities and the institutions of the state. Supporting horizontal (between communities and/or civil society) and vertical (upwards to different levels of government) networks are not mutually exclusive activities, and can be self-reinforcing. For example, local social networks can simultaneously provide the foundation for influence in local governance institutions and be significant for local learning and knowledge sharing. Importantly, a greater degree of connectivity between communities can provide increased opportunities for both activities.

Examples of working with networks in these ways are common in the case study chapters. In Bangladesh (Chapter 2), support for local people to form a community-based organization (CBO) gave access to local government officials and enabled adaptation activities to be undertaken in a political environment that routinely prevented collective action and blocked resource distribution. In Sudan (Chapter 8) a strong and representative local farmers' union was able to take on the role of lobbying the government on behalf of the community. The ability of communities to engage with governance and decision making can also improve resilience. For example, the input of local land users into natural resource management and enforcement institutions can have a stabilizing effect, reducing the likelihood of inappropriate or unsustainable land use changes (Chapin et al., 2006: 16,642; Adger, 2003: 398). In Nepal (Chapter 3) the formation of a CBO enabled the community to be represented in the local disaster preparedness network ensuring that their views on emergency response activities were considered.

The institutions, forums and relationships within social networks can also go some way towards converting climate knowledge from top-down, analytically based information to experience-based information. For example, Patt (2005) reports that farmers in Zimbabwe who attended workshops to learn about seasonal forecasts were significantly more likely to adapt farming methods in response to forecast information. In this example, the workshops were an attempt to build networking social capital. By engaging farmers in the process of developing a forecast, this approach attempted to make technical information amenable to experience-based reasoning (Patt, 2005: 12,624) and ensured that the forecasting information was credible, salient and legitimate to all stakeholders (Cash et al., 2006: 4). Establishing workshops is one approach to building networking capital to overcome the problems of communication and relevance of scientific information. Cash et al. recommend the use of 'boundary organizations' (2006: 4) – institutions that work as an intermediary between actors – while Patt (2008: 64) describes the forming of 'partnerships' between scientists and users. In each case an essential feature is the investment in social networks to build bridges between the knowledge systems and priorities of different communities, as achieved in Sri Lanka (Chapter 5) where links were established between community researchers and a seed conservation NGO, and in Kenya (Chapter 6) where the project brought together the community and meteorological office staff as active stakeholders.

While social networks define the access to and distribution of material as well as non-material resources, the experiences reported by Patt (2005) suggest that they have a particular role in promoting understanding of climate science and user needs. Moreover, the case study from Pakistan (Chapter 4) illustrates that such information-sharing and awareness-raising networks also help to build broader networks between individuals or groups with shared interests. However, it is important to recognize that effective communication of weather and climate information is not guaranteed by the development of networks alone. Experiences from seasonal forecasting also illustrate the challenges: Harrison et al. (2007b: 421) conclude that 'there has been a wealth of activity over recent years to promote the dissemination and uptake of [seasonal] forecasts. But, despite this progress, there are still few clear demonstrations of consistently-achievable value'. Significant problems remain with understanding the needs of farmers and other users of climate data, generating sustained institutional support and a favourable policy environment and, fundamentally, appropriate and effective communication of data for decision making. Understanding the needs of communities is essential: in Bangladesh (Chapter 2) it is poor infrastructure (roads and electricity in particular) that prevents forecasting information from reaching the community. Patt (2005: 12,623) reports how success in the use of forecasting in Brazil was undermined through the dissemination of one poor forecast. Elsewhere, Patt (2008: 64) notes an important barrier to communicating long-term climate change information: '[i]t is hard to develop effective partnerships between climatologists and users in the absence of a problem to be solved'. It is a common theme in the following chapters that it is hard to engage communities in discussions around the abstract concepts that underpin climate change. A notable exception is in Peru, where a rich history of observing and addressing climate variability ensured that the community engaged in lively discussions about climate change (Chapter 9).

Power

Social networks are significant in part because 'adaptation is mediated and interpreted through the lens of perceived opportunities' (Balstad, 2008: 173): expanding poor people's knowledge base and decision-making capabilities is thus an important function of networked institutions. However, the 'lens of perceived opportunities' can be refocused through the control of information and the motivations of stakeholders. Peterson et al. (1997) point out how 'management agencies often suppress scientific dissent in order to present a unified, "certain" front to the outside world, thereby consolidating the political power of the agency'. Such misrepresentation can easily gain traction owing to the attractive simplicity of the message (Balstad, 2008: 167) and may gain the active support of those who benefit from a particular view of the future. This scenario is pertinent to climate change, where the suppression of uncertainty legitimizes first-generation adaptation and may clear the way for commercial, technical fix solutions. For example, a diagnosis of a drying climate may facilitate the introduction of proprietary 'drought resistant' seed varieties, whereas an appreciation of the uncertainty associated with the prediction may instead lead to efforts to build the knowledge base of agricultural extension officers and ensure that a sustainable diversity of seed varieties is available. A similar scenario is evident in the Sri Lanka case study described in Chapter 5. The project was principally engaged in reawakening local knowledge of traditional rice seed varieties that have the potential to cope with the environmental changes now being experienced in Sri Lanka. While the country has around 2,000 traditional varieties, knowledge of their characteristics and benefits was eroded by the large-scale introduction of high-yielding, hybrid varieties that are now failing to cope with increasing salinity. The support provided for hybrids undermined local seed banks and indigenous knowledge, championing high yields and suppressing diversity to the long-term detriment of small farmers in Sri Lanka. Access to knowledge and information sources is therefore a necessary component of adaptive capacity, but insufficient without critical engagement. Network analysis provides an important tool for this process, enabling the chain of relationships between actors to be visualized and ensuring that the social, political and economic foundations of those relationships are assessed.

Different aspects of social networks may prove particularly important for adaptation. Moser suggests that greater attention be paid to understanding the 'social dynamics that underpin (motivate, facilitate, constrain) on-the-ground adaptation strategies and actions' in decision-making institutions, and specifically to address the 'value judgements and power dynamics embedded in adaptation decisions' (Moser, 2008: 188). Jennings (2008: 141) similarly draws attention to 'historically embedded and implicit power relations' and in particular notes the failure of indigenous or local environmental knowledge to penetrate bureaucratic ways of knowing. Ultimately, any local or NGO-led action, however positive, can be undone by a policy environment that is outright hostile to or simply lacks a focus on marginalized or poor communities. Moser summarizes her discussion of decision making as a call to 'stop hand waving about adaptive capacity and increase our understanding of, and our ability to use or create more effective governance structures to realize it' (Moser, 2008: 189). Networks of institutions, decision-making bodies and governance structures must, then, be subject to analysis (who represents whom, with what knowledge, with what motives?) if the translation of adaptive capacity into adaptation is to be pro-poor. While it may be possible to achieve the synergies between networks and the state idealized by Adger, in reality the dynamics of knowledge creation and decision making are politicized and must be recognized as such. Techniques can be employed to support and embed these aspects of social network analysis. Coupe et al. (2005) describe the use of farmers' juries in Zimbabwe to enable smallholder farmer learning on genetically modified crops through the production of witnesses for and against the technology. The same process enabled open and informed dialogue between farmers and senior government and Zimbabwe Farmers' Union (ZFU) officials, revealing major shortcomings in existing smallholder farmer policies and the deep dissatisfaction of farmers with how their interests were being represented by the ZFU. Alternatively, the use of on farm research in Sri Lanka, explored in Chapter 5, circumvents power holders by enabling farmers, rather than outsiders, to decide on the best crop varieties for their degraded lands.

Knowledge, vulnerability and adaptation: a coherent approach

The discussion up to this point has focused on concepts that are important to adaptation. In this section, the aim is to draw together the central themes of the chapter in a manner that highlights the relationships between the different approaches to and essential concepts of adaptation, summarized in Tables 1.2 and 1.3.

Table 1.2 Approaches to adaptation: different circumstances will demand a different blend of approaches

Approach	Comments
Vulnerability reduction	Vulnerability to climate change is assessed in reference to a particular hazard, for example vulnerability to flooding, and considers underlying human and environmental factors Vulnerability reduction targets a particular hazard, and should aim to be 'no regrets': meeting short-term needs while addressing potential climate change
Strengthening resilience	Defined as the ability to absorb shocks or ride out changes Reduces vulnerability to a wide range of hazards Supported by diversity of assets or livelihood strategies User input in decision making supports resilience by reducing the chance of damaging policy developments
Building adaptive capacity	Defined as the ability to shape, create or respond to change Strengthens resilience and reduces vulnerability to a wide range of hazards Amount, diversity and distribution of assets facilitates alternative strategies Requires information plus the capacity and opportunity to learn, experiment, innovate and make decisions

Table 1.3 Concepts essential to adaptation

Concept	Comments
Uncertainty	Adds complexity but is fundamental to understanding climate information and avoiding maladaptations Understanding and acting on uncertainty is hampered by the human desire for certainty and simplicity
Climate and forecasting information	Essential to adaptation decisions but requires knowledge of uncertainty and statistical nature of climate data, and the ability to interpret information expressed as probabilities Relevant focus and appropriate communication of climate and forecasting information is critical, but shaped by policy and influence
Social networks	Networks provide the opportunities for sharing knowledge, accessing resources and building influence Analysis draws attention to the social, political and economic context of relationships between actors Power is evident in the relationships that provide information, set policy and determine resource distribution

Social networks, knowledge and adaptation

Social networks are the glue between many of the elements of adaptation. They draw attention to the relationships between actors, and can be visualized as a web of connections that link diverse individuals and institutions, either directly or via other actors. For example, a household may be connected to other families in a village, a producers' association, a school and a political party. In turn, each of these actors has relationships with other parties that the household is able to access indirectly. The nature of these relationships will determine the household's knowledge of adaptation options, and its ability to send or receive goods, services and influence across the network. In this way, the analysis of social networks reveals a complex structure that governs the flow of material and non-material resources: the connections describe the access that different actors have to each other, while each link draws attention to the quality of relationships (including social, political, cultural and economic resonances and barriers) and the interests or motives of the different actors.

Policy measures can limit or extend the reach of networks through, for example, setting the rules and norms that govern institutions, and controlling the freedom to form new relationships. However, individuals are also important and the skills held by community members play a role in defining access to and activities in a network. Education in particular can underpin decision making, interpreting information and leadership potential. Literacy in particular can be important (education and literacy training appears in the Pakistan, Niger and Sudan case studies (Chapters 4, 7 and 8) as a development activity that supports adaptive capacity). NGOs can be a crucial part of a household or community's network, bringing knowledge, ideas, experiences and resources from the outside. Significantly, NGOs are also able to stimulate the growth of local networks by building contacts and relationships between previously unconnected actors, and by building capacity to interact with and access different actors and networks. Opportunities, lessons and resources for adaptation are accessed through social networks, often via indirect relationships (for example, through policy influence). While the ability to access and interpret climate change information is influenced by levels of education in particular, the information itself is mediated through social networks. The proportion of the currently available climate change knowledge that reaches a community will be defined by its social network, as will the complexity, accuracy and relevance of the information received. Working through networks can therefore improve access to climate knowledge in two ways: by extending the networks so that connections are built with holders of climate knowledge; or by exerting influence through the network to generate climate information that is more closely related to the lives and livelihood strategies of the community. Existing climate knowledge that is relevant but unknown to the community will (if it comes to pass) manifest itself as a (avoidable) shock; increasing the extent or relevance of climate knowledge thus reduces the possibility of these shocks.

The following characteristics are therefore critical, linking networks, knowledge and climate change adaptation:

- Policy (at multiple levels of governance, including local, national and intergovernmental) has an impact on social networks: for example by establishing institutions where actors can meet, or restricting free association.
- Different levels of policy play a role in the focus of climate change and forecasting knowledge by seeking to direct and commission scientific research (or by failing to do so).
- State adaptation and development policy affects communities and households, in terms of both asset distribution and the physical environment. The policy context will also determine the degree to which communities are subject to the actions of non-state actors (such as where multinational corporations constrain or control resource access).
- Ideally, communities will be able to influence policy development through their social network relationships: for example through lobbying, yielding changes in policy context and/or the focus of climate change and forecasting knowledge generation to take account of the needs of the poorest.
- Relationships between a community or household and the social network are bidirectional and multi-functional, bringing: 1) community or household activity in networks, carrying knowledge of the local context and needs, lessons from local experiences of adaptation, and the potential to learn; and 2) adaptation experiences, lessons and resources from outside to the community or household.

The adaptation space

Adaptation activities include measures to reduce vulnerability to climate change hazards, efforts to build adaptive capacity and actions to increase resilience. As noted above, these activities are interconnected in the sense that building adaptive capacity can increase resilience, while increases in both adaptive capacity and resilience can reduce vulnerability to a broad range of hazards. However, the key to understanding which adaptation activity – or blend of activities – is appropriate depends on where a community or household sits in what can be called the 'adaptation space'. This space is formed by the interrelationship of two key parameters, each of which needs to be considered when designing an adaptation strategy to a particular hazard: 1) vulnerability to the hazard; and 2) the state of knowledge related to a given climate hazard (existence, precision, uncertainty and trustworthiness).

Potential hazards can be identified through local knowledge of the areas in which livelihood strategies are linked to the environment and of the impact of previous climate variability and weather extremes. This process initializes an assessment of both parameters, identifying climate change predictions that are of significance to the community, and providing the first step in assessing vulnerability.

The first parameter is an assessment of starting-point vulnerability (discussed at length above) to the particular hazard or group of hazards in question. The second parameter can be described as the 'clarity' of climate change knowledge for a given hazard. The aim is to systematically assess climate knowledge so as to understand the extent to which it is suitable for driving the priorities of adaptation. Clarity can be taken as a compound appraisal of whether relevant knowledge exists, how precise that knowledge is, and its degree of uncertainty and trustworthiness. For instance, is the available climate or forecasting information focused on the regional, sub-regional, national or local level? Is there close agreement among model and scenario predictions or a wide range of opinion? And, do those that produced the knowledge, the reasons they provided it, and the time it was researched allow the user to trust its validity? Clearly, where there are multiple sources of climate knowledge available that agree with one another and give clear insight to the question in hand, the level of 'clarity' can be said to be high.

Plotting clarity against vulnerability on different axes creates a visual adaptation space that can aid understanding of the relationship between the different adaptation activities (Figure 1.1). The influence of clarity and vulnerability is felt in the following ways.

- *Low to high clarity.* Low clarity implies that the likelihood and/or nature of the hazard is poorly understood. Adaptation actions need to focus on improving knowledge clarity. This might involve improvements to climate modelling or in the ability of networks to access or demand more relevant climate knowledge from existing knowledge holders. Lack of understanding can also be guarded against through building adaptive capacity and resilience. The higher the clarity, the more adaptation actions can be informed by existing knowledge of vulnerability to the particular hazard.
- *Low to high vulnerability.* Low vulnerability demands little action in response to the hazard: development activities can continue as livelihoods are uninfluenced by climate change. High vulnerability will demand actions directed by the significant issues identified in the assessment of (starting-point) vulnerability.
- *Interdependence of clarity and vulnerability.* Vulnerability assessments to a particular hazard are only useful to the extent that there is clarity about the hazard. Therefore lower clarity demands improved climate knowledge.

Figure 1.1 also reveals how the possible adaptation options form a continuum that stretches across the two dimensions of vulnerability to, and clarity about, a particular hazard. Moreover, as a decision-support tool, it identifies the blend

30 UNDERSTANDING CLIMATE CHANGE ADAPTATION

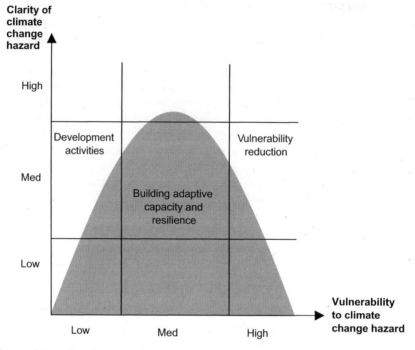

Figure 1.1 The adaptation space for a particular climate change hazard

of adaptation options appropriate to a particular situation and for a particular hazard based on an assessment of clarity and vulnerability.

Figure 1.1 illustrates the dependence of adaptation activities on vulnerability and clarity. A different mix of activities is defined for each grid square, broadly defined in proportion to the size of shaded and unshaded segments of the grid square. Plotting an assessment of low, medium or high vulnerability and clarity on the diagram draws attention to the necessity of employing a blend of adaptation activities for the particular circumstances. The following relationships are illustrated:

- Building adaptive capacity and resilience are significant responses unless clarity is high. This is because these activities address and insulate against the unexpected.
- Where vulnerability to a climate hazard is low, development activities can continue in the same manner as if climate change were not occurring. However, as clarity decreases, activities focused on building adaptive capacity and resilience become increasingly important in recognition of the inability to assess vulnerability accurately.
- Where vulnerability is high the adaptation response will focus on vulnerability reduction. While no-regrets approaches reduce the risk

of maladaptation, where there is less clarity about the hazard more investment is required in adaptive capacity and resilience, in recognition that the hazard may manifest itself differently from the manner anticipated by climate change predictions.
- Where clarity is high, adaptation actions move, as evidence of vulnerability increases, from existing development activities to include an increasing proportion of vulnerability reduction measures. Near the centre of this continuum development activities, which build broad coping ability by reducing poverty, merge into mechanisms for addressing specific climate challenges as guided by starting-point vulnerability.

The nature of the activities that result from an assessment of starting-point vulnerability will be highly context specific, and will vary depending on the category of hazard that is being addressed. Broadly, category 1 hazards (discrete recurrent events) are likely to benefit from disaster risk reduction strategies, while incremental changes (category 2) may require modifications or alternatives to, or diversification of, livelihood strategies. Hazards that alter environmental factors, such as changes to soil quality due to reduced rainfall, may require a blend of ecosystem and livelihood responses, such as measures to improve and conserve soils, and measures to capture reduced rainfall such as rainwater harvesting. Where vulnerability and clarity are both high, investment in infrastructure may be appropriate (for example, dam building in the face of potential glacial lake outburst, or sea wall provision to protect from slowly rising sea levels). However, in most cases a degree of uncertainty will exist, a fact that vulnerability reductions measures take account of by seeking no-regrets strategies that provide immediate benefit to the community. Category 3 hazards (shifts in climatic regimes) if predictable may demand a more extreme response: for example the abandonment of lands or particular livelihoods options. However, such extreme responses are not limited to category 3 hazards as incremental or recurrent climate changes may ultimately render marginal livelihood strategies untenable.

For each hazard category, the proximity of the impact will also be a crucial factor: hazards that are anticipated soon require an immediate response, most likely in the form of vulnerability reduction measures focused on livelihood impacts. Such hazards are also likely to have a high degree of clarity. This is reflected in the adaptation space through the association of high clarity with vulnerability reduction. Similarly, predictions further in the future are likely to be much more uncertain – both demanding and allowing time for the building of adaptive capacity.

The adaptation space is principally a conceptual model to aid understanding of the relationship between uncertainties and vulnerability. However, it may also operate as a practical tool, guiding analysis in a particular set of circumstances. As a practical tool it may be necessary to produce several versions for a particular community, each reflecting the clarity and vulnerability associated with a particular climate hazard. It is likely that in any given location there will be multiple hazards to consider (possibly of different categories), each with different uncertainty, likelihood and consequence. As time passes it will also be necessary to repeatedly revise both the position of the community or household within the adaptation space, and the particular hazard that is being considered. For example, the assessment of vulnerability may change as the hazard itself is moderated by clarity of climate knowledge. An assessment of high vulnerability to extreme rainfall may change to medium vulnerability to moderate rainfall if increased climate knowledge reveals that the extreme is very unlikely. Moreover, for a given class of hazard (say, change in rainfall) it is possible to move around the adaptation space in response to improved access to climate change knowledge, changes in adaptive capacity and resilience, and vulnerability reduction activities (for example, the formation of a watershed management organization and the construction of a community shelter to protect people and assets during flooding).

These observations draw attention to the dynamic nature of vulnerability and of adaptation. Vulnerability varies when the underlying human and environmental context changes. The dynamic nature of adaptation refers to changes in the nature of the optimum adaptation actions as time passes, illustrated by the shifting location of a community or household within the adaptation space. Two significant scenarios arise, denoted by the different trajectories in Figure 1.2. In the best case, an initial condition of medium or

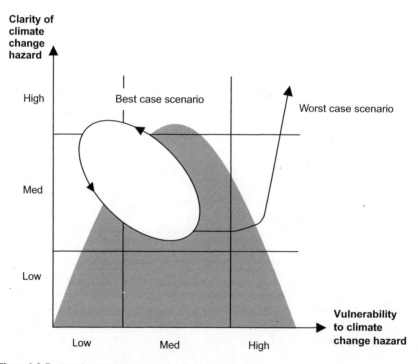

Figure 1.2 Best and worst case scenarios for adaptation

high vulnerability is met by an appropriate combination of adaptive capacity, resilience and vulnerability reduction measures, including an increase in climate knowledge and adaptation opportunities and experiences supported by social networks. These responses shift the community towards the top left grid square (higher clarity and lower vulnerability). Subsequent climate surprises – which are likely given the limitations of climate knowledge – will give rise to a reduction in clarity and increase in vulnerability. In the ideal case, existing and new adaptive capacity and resilience reduce the impact of the unexpected change and facilitate a swift return to the top left grid square. In the second scenario, however, the worst case emerges, in which vulnerability continues to increase over time, without an improvement in resilience or adaptive capacity. The failure to build contacts with climate knowledge holders means that clarity about climate change only emerges when the impacts start to be experienced. At this point swift vulnerability reduction measures are necessary, most likely focused on alleviating impacts and possibly leaving profound changes such as migration as the only option. In both cases, however, the starting point was one of medium clarity – a realistic scenario given the uncertain and probabilistic nature of climate predictions.

Culture, communities and adaptation practice

An appreciation of culture has an important role to play when implementing adaptation strategies and is at the centre of community-based adaptation. The need to adapt to climate change may pressure individuals and communities into changing livelihoods, lifestyles or patterns of behaviour, potentially challenging existing notions of culture. A series of questions arise: does, and if so how does, a shared culture provide, alter or limit the options for adaptation? How and why do individuals within communities respond to the prospect of changes to their lives and livelihoods? And importantly, what lessons emerge for those working to secure lives and livelihoods in the face of climate change? These questions are addressed in this section, in which culture is assumed to be 'the sum total of the material and spiritual activities of a given social group... a coherent and self-contained system of values and symbols [and] practices that a specific group reproduces over time and provides individuals with the signposts and meanings for behaviour' (Stavenhagen, 1998: 5).

Well-being and cultural context

Culture is important because of its significance to the well-being of individuals. The relationship between culture and well-being is rooted in the goals that an individual holds to be important. Goals, which may include life plans, relationships and ambitions, are themselves chosen and informed by culture, and the ability to work towards or achieve goals is critical to well-being (Raz, 1988: 291–311). The indicators of culture (known as social forms and encompassing shared beliefs, collectively held metaphors and folklore) are an ever-present factor influencing plans and decisions. In the same way, the goals held by an individual are informed by the prevailing social forms: the things that we consider to be important in life are not selected in a purely objective manner, but in reference to the culture that surrounds us. Choices made in response to the challenges of climate change will inevitably be made in this context.

For well-being, we have to see value in the activities that make up our goals, but our perception of value is determined by the cultural context; our actions 'only have meaning to us because they are identified as having significance by our culture, because they fit into some pattern of activities which can be culturally recognized as a way of leading one's life' (Kymlicka, 1989: 189). In the same way, similar activities will have a different significance depending on the prevailing attitudes and social practices (Raz, 1988: 311). Thus culture is central to establishing the significance of activities and dominates individual perceptions of well-being. These observations are central to how individuals and communities respond to the prospect of change: 'freedom of choice is dependent on social practices, cultural meanings, and a shared language... the context of individual choice is the range of options passed down to us by our culture' (Kymlicka, 1995: 126). This suggests that adaptation options are limited: not every available option will resonate with local social forms. For example, to suggest the use of human waste in composting would be met with abhorrence in some cultures, yet is commonplace in others. However, a more positive reading of this analysis suggests a productive role for culture, in which the social forms that make up culture define opportunities for adaptation, and suggests that for success, changes should be rooted in or build on local culture.

The above analysis also suggests that individuals and communities will respond to the prospect of change differently depending on how and why change emerges. Notably, the importance of community and cultural identity can alter in different circumstances, a fact that becomes evident when identity politics spills over into violent conflict. The subjective sense of group identity may be reinforced among communities that are disenfranchised, face competition for resources, or are threatened. However, this observation is not the same as suggesting that culture is necessarily resistant to change, and it is erroneous to paint cultures as fixed, unchanging entities. Rather, it suggests that changes that are perceived as a threat to culture are likely to be resisted, closing down opportunities for developing the local context of choice. For example, women's freedom to work outside the home is strongly constrained in many countries, meaning that proposing small-scale trading as a livelihood opportunity for women is likely to be met with resistance. Moreover, the proposal itself, showing a lack of sensitivity to social forms, could offend and thereby inhibit future dialogue. The process of engagement is therefore critical. Attempts to impose changes from outside should be avoided, and instead the full involvement of communities in the process of adaptation should be promoted: in short, change should be developed from within cultures rather than from without.

Culture and adaptation

If those working with communities seek to identify local problems and locally appropriate solutions, activities should naturally build on social forms (working with cultures) and provide an opportunity to extend the local context of choice (effecting change from within, through dialogue and developing a local understanding of the challenges of climate change). Rooting the process of adaptation in communities allows important communal practices and collectively held metaphors or sayings to be identified and used to facilitate change from within, rather than attempting to force change from without. Cultures that lack a tradition or history of adaptation (to climate or other environmental challenges) require an approach that builds from existing social forms and is sympathetic to local notions of well-being. This may be more or less of a challenge depending on how radical a transformation is required and whether the existing cultural context of choice is narrowly defined or deeply entrenched. The effectiveness of engagement with and appreciation of culture is evidenced in the case study chapters. For example, in Pakistan (Chapter 4) local social forms were capitalized on to communicate climate change and natural resource management messages, with traditional institutions and festivals revived to mobilize community action. These interventions were found to be more successful than attempts to form community-based organizations, which were alien to the local culture. In the same context, gender roles are deeply rooted, to the point that public social interaction between women and men is not acceptable. To be acceptable, project work was divided between exclusively female and male groups. In Niger (Chapter 7), work with a nomadic community in the Azawak region introduces change while retaining a key societal marker: the approach uses 'fixation points' that allow the communities to retain their migratory traditions while offering access to a diverse range of services at dry season congregation points. More generally, as investment in meteorology yields ever more climate and forecasting information, innovative and culturally responsive approaches to communication will become increasingly important.

The role of culture outlined here indicates the need for a nuanced approach to adaptation that is grounded in a highly developed appreciation of local social dynamics. Community-based adaptation is well placed to offer this as its methodology seeks to work with communities to define problems and solutions. It may be that situations are encountered 'where "local culture" is oppressive to certain people' and may rob the most vulnerable within a group of a voice (Cleaver, 2001: 47). However, an understanding of culture can also help to transcend a simplistic view of power relations as oppressive, and instead point to the complex role that power holders play in well-being and the entry points that they offer into communities (Ensor, 2005: 266). Effective action will require a line to be walked between the extreme 'we know best' and 'they know best' positions (Cleaver, 2001: 47) by seeking to work within cultures to build and develop dialogue on the challenges of climate change. As Twigg (2007) notes in the context of disaster-resilient communities, success will see the chosen approach to adaptation become a shared community value or attitude. However, reaching this point also requires an enabling environment – a 'political, social and cultural environment that encourages freedom of thought and expression, and stimulates inquiry and debate' (Twigg, 2007: 26). Engendering such an environment requires engagement with the interplay of politics, policy and power at multiple levels, and is a key challenge in the process of moving community-based adaptation beyond isolated development projects and into institutionalized policy frameworks. It is a topic that is returned to in the conclusion to this book.

Structure of the book

This chapter has set out to frame adaptation to climate change within the context of the inherent uncertainty of climate change predictions. The focus is on resource-poor people and addressing the additional vulnerabilities that climate change causes. Different levels of clarity about the implications of climate change in different regions imply a continuum of approaches to adaptation, from a focus on development as usual when climate change is not imminent and not expected to impact severely, to a focus on vulnerability and disaster risk reduction where there is greater certainty that climate change will cause severe impacts in the near future. Actions to increase adaptive capacity and resilience are important where there is some uncertainty about the future and should therefore form an important component of most adaptation interventions.

The ingredients of a programme that will assist communities to adapt are explored through eight case studies in the following chapters. Four of these cover work in South Asia, drawn from the first adaptation project implemented by Practical Action. These cover farming communities in remote areas experiencing a range of climate challenges, from more frequent heavy rainstorms, more intense monsoon flooding, to longer dry seasons and a new phenomenon of winter cold spells. The work in three of these countries was carried out directly by Practical Action programme staff from the local country office, while the project in Pakistan was implemented by a long-standing partner of Practical Action, the Rural Development Policy Institute (RDPI). A fifth case study explores work in the Andes of Peru, where Practical Action has been working for 20 years with indigenous communities surviving very harsh conditions at high altitudes. These communities are now experiencing severe hailstorms and long periods of frost, coupled with very low rainfall and intense sunshine. Of the three case studies from Africa, all relate to communities facing reduced rainfall. JEMED, a partner of Tearfund, has been working to help pastoralists in Niger facing loss of traditional rangelands to adapt to a drying climate; in semi-arid central Kenya, Centre for Science and Technology Innovations (CSTI) worked with small-scale farmers facing drought and natural resource degradation. The case study from Sudan relates to a resilience-building project by the NGO SOS Sahel with an agro-pastoralist community that has faced several extreme droughts in recent decades. This chapter has a slightly different focus, in that it presents the results of a research study review of the project to assess its work in building adaptive capacity.

Each chapter has been prepared with the participation of one of the team members that undertook the project work. An account is given of the project context, existing vulnerabilities and the particular challenges that climate change (or increased climate variability) is currently bringing. The process of understanding existing capacities and knowledge to build adaptive capacity and reduce vulnerability are described in some detail, as are methods to extend the communities' social networks, and approaches to strengthening livelihoods in the face of climate change. In some cases raising awareness of climate change forms a major element, while in others technology and/or skill development are the main focus. A section on lessons learnt is intended to offer a frank discussion of what worked well and the weaknesses that emerged in each approach. The conclusion to each chapter draws out key elements for adaptation in the particular context, with reference to the concepts developed in the Introduction.

Finally, conclusions are presented in the last chapter. This explores the components of an adaptation project in light of the framework for adaptation set out in this Introduction, examining the relationship between reducing vulnerability to climate change, building adaptive capacity and strengthening resilience. The role of climate information in projects is discussed, along with the implications of predicating community-based adaptation on existing vulnerabilities. The second half of the conclusion addresses the challenges of scaling up adaptation projects, noting in particular the need to address existing policies, politics and power relationships if communities are to be empowered to identify and develop appropriate responses to the challenges of climate change.

Community-based adaptation remains a new concept, for which good practice must be developed and shared widely as a matter of urgency. It is hoped that this book will be a useful contribution to academic debate, will inform policy at the national and international levels on how to support adaptation, and will help the many development organizations and practitioners throughout the world working to help poor people in the face of climate change.

Endnotes

1. The IPCC Chairman, Rajendra Pachauri (2006: 4), has stated that the role of the IPCC is to 'review and assess policy relevant research; i.e. not be policy prescriptive, but policy relevant... relevance has to be based on our

perception of the decision-making framework and the kinds of issues that become part of policy'.
2. The models were all run using the IPCC's A1B scenario, in which the world economy is assumed to expand and global governance emerge. The 'B' refers to 'balanced' progress in supply and end use technological change (compared with, for example, the A1T scenario in which a rapid transition to a non-fossil fuel future is assumed).

CHAPTER 2

River erosion and flooding in northern Bangladesh

Prepared with K.M. Mizanur Rahman, formerly Programme Manager, Practical Action Bangladesh

Abstract

> The char islands in the Gaibandha District of northern Bangladesh are created and eroded by the flow of the Tista and Brahmaputra rivers. Those living on the islands or banks of the rivers are extremely vulnerable to changes in river flow that rob them of their cultivable land or homes. This chapter focuses on a project aimed at 2,000 households that were living in temporary locations on riverbanks, embankments or public places having been displaced by flash flooding events that are becoming increasingly frequent. The project worked to build social networks within and outside the community, facilitating awareness raising, training, community participation in local government institutions and the development of livelihood, income generating and flood proofing technologies.

Introduction

Bangladesh is one of the most densely populated and low-lying countries in the world, prone to floods, cyclones, storm surges, tornadoes, earthquakes and droughts. While tropical cyclones and the accompanying tidal surges produce the highest number of casualties, it is floods that induce the most widespread and prolonged damage to the livelihoods of the poorest. The flood plain of Bangladesh is fed by the catchments of the Ganges, Brahmaputra and Meghna rivers in India, Nepal, Burma and Bhutan, bringing 1.2 trillion (1.2×10^{12}) cubic metres of water annually. While the catchment area totals in excess of 1.5 million square kilometres, in Bangladesh the run-off is channelled through a country covering an area of only 144,000 km². Climate change presents multiple dangers in this context. River flooding and riverbank erosion are becoming more severe as rainfall intensity and glacial run-off rates change, while in the southern flatlands rising sea level threatens millions of people with flooding and long term salinization. Thirty years of temperature and rainfall data from the Bangladesh Meteorological Department reveals that the annual number of days without rainfall is increasing, but that the total annual rainfall is unchanged. This trend towards heavy rain over a short duration is bringing increased river erosion and droughts, and is anticipated to continue under current climate predictions.

The adaptation project reported on in this chapter was implemented in Gaibandha District, one of the most vulnerable areas in Bangladesh. Gaibandha District is located in northern Bangladesh at the confluence of the Tista and Brahmaputra rivers. It is an area particularly vulnerable to floods and riverbank erosion. The annual monsoon flows cause the meandering rivers to erode the banks, while elsewhere silt may be deposited to create new land and riverine islands, known as *chars*, in the middle of the river. People living in these areas face loss of cultivable land and even their homes, sometimes five or more times in their lives. While the creation of new land offers new opportunities for settlement, this land is sandy and less fertile than the lost land, and can also be subject to competing claims. Often, those who owned *char* land prior to the most recent cycle of erosion are driven out by the local elites when a new *char* is about to be formed. The more powerful elites then grab the newly accreted land. However, when tenure remains with the old settlers, they may try to claim their land, fuelling local conflict. In recent times, local elites have made it increasingly difficult for the poor people to enjoy the benefits of newly accreted *char* land.

As recently as 20 to 30 years ago most livelihoods in the area were based around fishing and farming, with enough land available for households to cultivate a variety of crops. However, frequent flooding and riverbank erosion have severely degraded the assets of local people, who have lost land, houses and livestock. Farmers have suffered multiple setbacks: the loss of land due to riverbank erosion, increased drought reducing the water available for crops, and flooding that can wash away entire fields of crops and leave the land impossible to cultivate. Fish cultivation also suffers during floods, as fish held in enclosed ponds can be washed out to the open water body. The market price of fish also drops during floods, reducing the income available from fishing. Where most of the community were until recently relatively wealthy farmers, the impact of flooding and drought has led to destitution for many. Most now rely on subsistence or very small scale livelihood activities, or migration to the cities to undertake day labouring or rickshaw pulling for an income during floods. Women and children are particularly vulnerable as they lack alternative livelihoods and institutional support. Many women work outside the home as day labourers or as maid servants. The marginalization of the communities is illustrated by the lack of government support reaching the Gaibandha farmers: the demand for extension services at the *upazila* level (sub-district, around 100 villages) far outstrips the number of government officers; posts created to meet the additional need remain unfilled, and funding is so low that those officers that do exist are unable to undertake field visits.

With the intention of providing assistance in addressing the climate challenges facing the Gaibandha District communities, the major objectives of the project described in this chapter were:

- to strengthen the capacity of communities, government and non-government supporting institutions to prepare and respond effectively to future climate-induced emergencies;
- to develop and promote practical interventions to strengthen livelihoods and natural resource assets;
- to promote the engagement of vulnerable communities in decision-making processes on climate-related adaptation strategies in order to influence policy change and increase self-sufficiency.

The project aimed to reach 10,000 vulnerable men, women and children. At the initial stage, the project staff conducted intensive discussions with the partner NGOs, local elected public representatives and local government service providers to select the vulnerable villages. The selected villages were recognized to be prone to flood, riverbank erosion, cold periods and drought, while the 10,000 beneficiaries comprised 2,000 households that were frequently displaced by flooding and living in temporary locations on riverbanks, embankments or public places. The members of these households exist in temporary shelters and own very little or no land. The remainder of this chapter describes Practical Action Bangladesh's three-year engagement with these communities.

Community-based adaptation in the northern *charlands*

The project was initiated with a series of discussions and surveys that provided a baseline understanding of the community's knowledge and awareness of weather, climate and climate change; analysed local livelihoods; and identified the assets relied on, owned or accessed by the community. While the climate and weather data provided the basis for building an awareness-raising strategy, the livelihood and assets studies showed the vulnerabilities and strengths of existing livelihood practices in relation to potential climate change impacts. The baseline study provided sufficient descriptive data for project design and provided the backdrop to participatory technology development and training.

Assessing climate knowledge and awareness

The project commenced with an analysis of how local people accessed weather and climate information, revealing use of both traditional and modern knowledge systems. Most of the elderly in the community stated that they depend on traditional knowledge in order to predict weather conditions. To do so, they observe a range of natural indicators:

- Ants carrying their eggs with their mouths and moving upwards to a safe place, crabs rising up into the community's houses, and earthworms emerging from their holes all indicate that flood will occur.
- A storm or cyclone is coming if the *sarashi* (a small insect) bites people.
- Sultry weather indicates heavy rainfall accompanied by a storm.
- If heavy cloud moves from south to north then flood will occur within two weeks.
- If heavy clouds appear in the north-western sky, high rainfall is expected.
- Wind from the south indicates that the river will be strong.
- When *dhaincha* (a local species of forage plant) grows more than usual there is the possibility of flood.

Communities continue to make use of traditional forecasting knowledge, the more so in remote areas. However, in recent times traditional methods have become less reliable and, generally, the communities rely on a combination of traditional and modern sources, comprising:

- *Observation.* As the farmers and fishermen live very close to the rivers, observation of the river level is the most important source of information. Many community members report having to move their valuables to high ground during floods in 1998 and 2004 as a result of watching river levels. However, observation is insufficient to provide an early warning mechanism against flash flooding.
- *Word of mouth.* Visits to markets or the houses of relatives provide an opportunity for community members to discuss issues, including exchanging weather information. Both traditional and modern forecasting information is shared in this way, with literate friends or relatives passing on reports from written sources.
- *Radio and television.* Weather-related information is announced regularly through radio and television, including bulletins that trigger emergency preparations. Some community members have a radio; those that do will usually share information with their neighbours. Television provides a more vivid indication of conditions by showing pictures of the river level, but is restricted to wealthier households or public television sets such as in the market, and then only when electricity is available.
- *Newspapers.* Newspapers are generally not available in the villages because of poor communication networks; however, some educated community members may access newspapers at nearby markets or towns. This information will usually be shared with other community members.

The formal sources provide national rather than local weather information. For example, while seasonal planting information may be disseminated, or flood information provided by the Flood Forecasting Centre, the focus does not reflect the needs of the communities in Gaibandha District. Moreover, a lack of infrastructure to connect isolated communities to the outside world (including accessible roads and electricity) hampers the delivery of information to communities. There is a need both to extend the reach of formal information bulletins, for example through the use of notice boards, mobile phone text messaging, transmission of information by bicycle or rickshaw, and to validate local observations of natural phenomena through calibration with scientific observations.

The results of 16 focus group discussions at the start of the project revealed that few had heard of climate change. While there was a common perception that weather patterns had changed over the last 20 to 30 years, it was felt that this was natural or an act of God. Participants in the discussions described problems such as land loss and inadequate food supplies, and blamed these on natural disasters, particularly flood, river erosion and cyclones. Changes such as extended periods of hot or cold days, or dense fog, had been noted, but the community did not relate any of the observed changes in weather to future or long-term climate change. Similarly, stakeholder discussions at the inception of the project revealed that those working in extension institutions were not aware of climate change or the likely impact on local farming and fishing communities.

Social networks

A central pillar of the project was to build networks within the community and with outside stakeholders. Community volunteers and youth volunteers were recruited from within the locality and were the focus of training on climate change issues. These groups were subsequently charged with awareness raising in the community and management tasks within the project. Community volunteers were selected following group discussions involving civil society, religious leaders, schoolteachers, members of the *Union Parishad* (local government, covering around 10 villages), local elites, community leaders and local NGO personnel. Two hundred volunteers aged between 18 and 40 years were selected (100 women and 100 men) each of whom was required to demonstrate: 1) permanent residence in the community; 2) ability to read documents in their local language (formally educated to at least level VI); and 3) positive attitude and commitment to voluntary community work.

Teachers were involved in selecting youth volunteers from among local school students. As with the community volunteers, the students were required to be residents in the area and be studying above class VI. The students had to be willing to work during emergencies and their involvement in the project had to have the support of their family. One hundred and fifty youth volunteers were selected, with equal numbers female and male, and each received training on the climate change challenges facing Bangladesh.

The community and youth volunteers were engaged in awareness-raising activities in the locality, while the community volunteers were also involved in preparing a community disaster preparedness plan. Community members were asked to identify the types of disaster that occur in different months, the associated impacts (for example, loss of crop, land or other assets) and the methods commonly employed to avoid the worst of the impacts. With a view to improving planning and coordination, the disaster plan allocates roles to the community volunteers before, during and after a disaster, and charges each volunteer with communicating the plan and providing support to 10 households.

The community volunteers also established two community-based organizations (CBOs) from within their membership. Known as 'early warning committees', they were created in an effort to build relationships outside the community. In particular, *upazila* officers responsible for agriculture, fisheries and health were identified by the community volunteers as key contacts who were failing to provide services to the community. Membership of a CBO, if registered with the *Upazila* Social Welfare Department, enables community members to become involved in local government committees. In this way, the formation of the CBOs helped to build important relationships between the community, government and non-governmental organizations. These were sufficient to enable the community to work on adaptation activities in a local political environment that routinely inhibits collective action and blocks resource distribution. Work undertaken included the construction of small roads, digging of culverts and tree planting. The CBOs were also empowered to participate in disaster preparedness meetings at the *Union Parishad* – ensuring that the community was able to engage in and provide feedback on emergency response activities during disasters. To inform and facilitate their work, the committees attended two training sessions. In the first, technical training was provided to build skills; this was facilitated by *upazila* government officers. In the second training session, information on the impacts of climate change was offered, facilitated by project staff and an external expert. One notable outcome of the training process was that the technical training was much better received by the community members as it was perceived to offer help directly relevant to current concerns, unlike the more abstract information about national or global changes in climate.

The project carried out a number of training events on climate-related issues to build capacity and develop skills among the community and youth volunteers, and in different vulnerable groups, government institutions, local elected bodies and local NGOs on climate change issues. Table 2.1 documents the range of courses and the number of beneficiaries.

Awareness raising

Along with formal training courses, the project also initiated activities intended to raise awareness of climate change within the community and among external stakeholders as a step in gaining support for adaptation activities. The assessment of local and institutional knowledge carried out at the start of the project informed the communication strategy design process, leading to

Table 2.1 Courses held during the community-based adaptation project

Course	Number of Courses	Number of Participants
Training and refresher courses for community volunteers on climate change issues	24	200
Training and refresher courses for youth volunteers on climate change issues	16	142
Training for teachers/civil society/CBOs on climate change issues	2	48
Awareness raising and training courses for local elected bodies on climate change	2	49
Training and refresher courses on early warning systems	4	50
Training of farmers on livestock rearing during disasters	2	50

activities focused on the causes and impacts of climate change using a variety of media, as detailed below and summarized in Table 2.2.

- *Publicity materials.* Brochures, leaflets, placards, billboards and posters were developed by the project staff and a consultant. The materials were used to disseminate information to the community and external stakeholders. They highlighted the impacts of climate change, how to prepare for different disasters, gave details of early warning systems and explained how the community can best respond during disasters.
- *Cultural shows.* Local cultural groups were engaged to disseminate messages to the communities though cultural programmes. The groups developed songs highlighting climate-related issues and organized public shows. This was a popular and successful approach, particularly in rural settings where thousands of people would gather and hear messages through songs that were in local languages and in tune with the local culture.
- *Debates.* To create awareness among the school students on climate change issues, a debate programme was arranged with topics relating climate change impacts to local livelihoods.

Table 2.2 Awareness-raising activities undertaken during the project

Types of material and event	Number of items or events
Banners	20
Posters	1,200
Signboards	8
Rickshaw tin plates	180
Billboards	4
Video documentary	1
Debates	4
Art competition for school students	1
Essay competition for school students	4
Rallies	4
Cultural shows	4

- *Art and essay competitions.* Climate change-related art and essay writing competitions were arranged for school students in the area. This process both enhanced awareness and improved the students' knowledge of climate change.
- *Rickshaw adverts.* To create mass awareness on climate change issues, climate messages were written on tin plates and distributed to rickshaw pullers. Fixing the plates to the back of the rickshaws brought climate information to a wide audience (as evidenced by follow-up interviews outside the immediate project area) and was subsequently adopted as a communication method by other government and non-government organizations (for example, public health information was distributed in this way).
- *Rallies and celebrations.* Community awareness was raised through the observance of events including World Environment Day (organized by the UN annually on 5 June, to encourage awareness of environmental issues globally) and International Day for Disaster Reduction (13 October, organized by ISDR, International Strategy for Disaster Reduction). Several rallies were organized by the community and youth volunteers, with an average of 500 people attending. The decision to hold a rally was taken by the community and youth volunteers who subsequently shared information about this event with their community (for example, each community volunteer relayed information to the 10 households for which they are responsible).
- *Video.* Video footage was recorded in the project area to capture examples of the impacts of climate change and demonstrate adaptation initiatives that had been undertaken. The video was presented to local and international policy makers, donors and researchers.

Technology development

During the second year of the project, community-led identification and prioritization of natural resource management options and technologies took place. Participatory discussions and subsequent training focused on increasing resilience through the adoption of new and improved technologies for crop and vegetable cultivation or livestock raising. The likely effects of climate change were considered in the decision-making process, with a particular focus on coping with increased flooding. The needs, skills, assets and capacities of the community were assessed through household surveys, participatory action plan development (PAPD, an approach that combines consensus building and participatory rural appraisal – PRA – principles) and consultation meetings. Training and input support was then tailored to the particular profile of household members.

The technologies identified and developed during this project include:

- *Portable cooking stoves.* Cooking food is a major problem during floods. Traditionally households have used fixed stoves that are constructed at ground level and cannot be moved during floods: as a result a movable, improved cooking stove proved a popular development. The stoves are very cheap to build, consume less fuel and produce less smoke, thereby improving women's health in particular.
- *Tree nurseries.* For those with land available, cultivating flood-tolerant tree species provided an alternative income opportunity. Seedlings are ready to be sold on within three months of planting.
- *Floating vegetable gardens.* Floating gardens allow crops to be grown during flooding periods, increasing the availability of food at household and local level. Vegetables are not normally available during periods of flooding due to inundation of cultivatable land, while many farmers have become landless due to riverbank erosion. A floating garden comprises a raft (approximately 3 metres by 1 metre) constructed from freely available water hyacinth. On the top of the raft, a soil layer is introduced, within which vegetable seeds are planted. Summer and winter vegetables such as gourd, okra and greens can be cultivated.
- *Homestead vegetable gardening.* As rain or flood water recedes appropriate vegetables are cultivated for family consumption. Any surplus can be sold, providing an additional source of income. A small piece of homestead land is sufficient for vegetable cultivation.
- *Caged fish cultivation.* Unlike the wealthy, poor community members lack a pond in which to cultivate fish, relying instead on using nets to catch fish in the open river. However, by using a cage, made from steel frame and net, fish can be cultivated in the rivers. Moreover, during floods the cages can be kept in front of the house, helping families to fulfil their protein needs (two or three cages are required for a typical family at a cost of around £50 (US$74, exchange rate £1 = $1.55, 18 December 2008) per cage).
- *Duck rearing.* Ducks are relatively easy to rear and are well suited to coping with floods. A household can strengthen their livelihood by selling eggs, while during floods they are able to supplement their protein intake by consuming the eggs and ducks.
- *Raised flood-proof houses.* Plinths of soil and cement are built to raise the floors of the houses above the flood level; reinforced concrete T-shaped pillars form the corner posts, and woven bamboo matting forms the walls. In the event of extreme flooding, or river erosion, all but the plinth can be moved to a new location. This is a high cost activity (around £150 per house) undertaken only with the poorest households.
- *Short duration rice varieties.* Cultivation of fast-maturing rice varieties enables harvests to be brought in before the monsoon. Yields of these varieties, developed by the Bangladesh Rice Research Institute, were also reported to be greater than previously grown varieties.

- *Improved goat breeding.* Goat breeding is being carried out to raise the productivity of local disease-resistant animals. Goats are distributed as part of a scheme in which each family hands on one of the kids bred from their female goat to another family. An innovative community insurance scheme was also applied in which death or failure to conceive is compensated through the provision of a kid bought from the communally held pool of insurance contributions.
- *Elevated tube-well.* The availability of drinking water is a major problem during flooding. Elevated tube-well technology uses bricks, sand and cement to raise the top of the well above the flood level, thereby preserving the water source. This is also a high cost technology at around £200 per installation.
- *Elevated latrine.* People usually face enormous difficulties during floods as toilets disappear below the water level. Elevated latrines remove this problem, improving hygiene during floods (but at a cost of around £100).

Technical training was necessary to build skills and confidence in the above technologies. Groups of community members were trained together by an invited expert such as the *upazila* fisheries officer (fish cultivation training) or agriculture officer (crop cultivation training). The training was provided in a practical format and was well received; traditional classroom-based sessions did not prove popular, particularly among the majority of participants who had little formal education. In some cases training also included visits to locations where successful examples of the technologies could be observed. In these instances the community members were readily able to adapt the experiences of others to their own circumstances. For example, farmers from the project area visited southern Bangladesh in which floating garden technology has been successful for raising seedlings. The Gaibandha farmers recognized that the existing raft design was too large for their river conditions, and felt that there was no local market for seedlings. However, they saw the rafts as an opportunity to help secure their livelihoods and so adapted it for vegetable cultivation. Skill development training then continued in Gaibandha, with farmers receiving training on the construction of the floating beds and families trained in cultivating vegetables. Each household was supported with the provision of seeds of 10 high-yielding varieties of fast-growing vegetables and groundnut seeds. Training also focused on methods to grow crops throughout the year, protecting against plant disease and insect attack using organic methods (such as home-made botanical insecticide), soil enrichment and compost preparation.

In total 1,800 community members received skill development training across all of the technologies rolled out in the project. While most of the technologies have been accepted by the community, some of those who attended training workshops have not yet applied their new skills, preferring to wait and judge the success of others.

Resource implications

The project was implemented by four people, two of whom were employed by Practical Action Bangladesh and two from partner organizations. The provision of hardware (for example, seeds, tools and housing supports) amounted to £19,730. Including training, workshops, staffing costs and overheads (including transportation) the total project expenditure reached £93,790. With a total of 10,000 beneficiaries across 2,000 households, the overall cost per household was £47.

Lessons and challenges

The starting point for the awareness raising, capacity building and technological activities undertaken in this project was a series of surveys designed to identify livelihood assets, needs and risks. While successful in enabling the design of productive interventions, project staff identified weaknesses in the process that could be addressed in future work. The baseline survey was conducted using a single questionnaire to capture capacity, assets, knowledge and awareness. However, it became clear that different groups in the same project area may require more specific or tailored questions to address similar issues: farmers and fishermen, for example, experience very different impacts during and after flooding events and these will not necessarily be captured via generic questions. Moreover, timing was critical in gaining access to project participants: dry periods in the project area are the times of highest livelihood activity and therefore the least appropriate for conducting intensive surveys.

A more general problem emerges if too restricted a view of the climate change problem underpins the survey questions and the subsequent analysis. A focus on livelihood assets and natural resources is understandable when attempting to establish the impact of environmental change on communities that are dependent on local natural resources. While significant, the immediacy of the impact of climate change on such communities should not distract attention from important underlying social, cultural and political factors. Gender (for example, differences in climate change impacts between men and women), culture (creating barriers to access to shelter or early warning) and politics (influencing resource distribution) all need to be assessed in background surveys if the context within which climate change and adaptation take place is to be properly understood. The increase in complexity that this approach demands suggests that a multidisciplinary project team or advisory group is necessary in designing and evaluating project activities, while community-based monitoring and evaluation may help to provide a focus on the breadth of issues that are significant in adaptation.

Awareness raising and training

The information gathered during the assessment phase of the project established the level of local and institutional knowledge of climate change and provided a clear view of the awareness raising required among different groups. This in turn formed the basis for the development of a communication and awareness-raising strategy. In particular, the baseline information suggested that both formal and informal methods would be necessary if all members of a community with varying levels of formal education were to be reached. This insight proved valuable and gave rise to different but equally successful strategies. To reach the broadest number of people, relatively simple climate messages were distributed on billboards and signs that were placed in communal areas. Similarly, the tin plates distributed to rickshaw pullers contributed to the wide reach of climate information, as did posters, leaflets and cultural events. Art competitions, essay writing and debates, however, were targeted at school students and their teachers. The youth volunteers were responsible for organizing the art and essay events in their schools, while teachers were instrumental in running the debates. Teachers and project staff selected topics relating the impact of climate change to the lives and livelihoods of the communities, successfully generating detailed awareness among both teachers and students. Training and workshop events were arranged to increase participants' understanding and awareness of climate change and, in the case of government service providers, to understand how – and whether – they intended to incorporate climate change issues in their development activities. Updated information on climate change issues was shared with the participants to improve their knowledge, emphasizing the need to consider climate change a priority issue in development planning. On a more general level, the process of developing a community-based disaster risk reduction plan assisted with broad awareness raising about the threats faced by the community, and improved local knowledge of specific approaches to risk reduction.

Despite the success of these targeted strategies, the complex messages that are inherent in discussions of the global causes and impacts of climate change remained extremely difficult to communicate to the local community. While the project organizers felt it was important to provide communities with a sense of where responsibility for climate change lies, this proved to be a difficult concept to communicate in a community that understood environmental change to be natural or the will of God. Moreover, training sessions faced the problem of finding locally relevant climate change information to satisfy the needs of participants who were focused on the challenges in their own communities. While this problem is insurmountable until local-level climate information becomes available, the immediate needs expressed by the community could be more appropriately met through local weather and seasonal forecast information. For example, the interests of farmers would be met if the *upazila* agriculture officer was able to translate seasonal rainfall trend information into recommendations for crop cultivation. While this is not climate information in the sense of providing long-term information, such an approach would allow adaptation to short-term variability and, if successful, build confidence in and awareness of scientific forecasts.

As noted, training sessions based around informal demonstrations and practical workshops were particularly appropriate when working with communities that have little experience of formal education. These sessions were further improved through the use of local trainers with knowledge of the local language and culture. The recruitment of these trainers was a significant boost, and was facilitated by partner NGOs who were able to make use of their contacts in the project area. Technical training for government officers and local NGO personnel faced different problems: while communication was less difficult, the application of the skills acquired in workshops was inhibited by a lack of resource and policy commitment to working on adaptation in remote communities.

Technologies

All the technologies introduced during this project were developed with and to meet the needs of the communities. They provide alternative livelihood opportunities during floods and increase the ability of the communities to cope with flood water inundation. Many were very successful when introduced: cultivating vegetables during flooding increased the availability of food at the household and local level; promotion of improved cooking stoves increased fuel efficiency, improved respiratory health and provided families with the ability to prepare food even during periods of flood; hygiene at the household and community level was improved by the introduction of elevated latrines; and the use of cage cultivation for fish proved very popular among the poorer community members. However, the cost of many of these technologies has the potential to inhibit their broader use. The high initial cost for the steel fish cages prevented many from adopting the approach, so during the course of the project a bamboo version of the cages was developed, significantly reducing the start-up cost. In other cases cheaper alternatives are less readily available: flood-resistant housing and elevated water wells and latrines proved to be excellent technologies but the project could only fund a few example installations. After the project was completed the community lacked access to those with the financial resources and technical skills to reproduce these examples. At the other end of the financial scale, floating gardens were very popular for the production of short-duration crops. However, in some areas this technology proved inappropriate because of the scarcity of water and water hyacinth in the dry season when the floating beds need to be prepared.

All of the technologies introduced during the project would be suitable for wider dissemination. However, discussions with the community also revealed an alternative priority: that of finding a permanent solution to the year-on-year loss of land to flooding and erosion. Protection such as that provided through the introduction of embankments would be possible, but the costs would be huge. Difficulties remain, however, with scaling up even the technologies demonstrated in this project. Government policy and resource support will be crucial to ensuring impact at scale, so awareness of the likely impacts of climate change, the need for adaptation and the benefits of scaling up the demonstrated technologies need to be shared with government service providers and policy makers in various ways, including through demonstrations at the project sites.

Conclusion

A central theme of this project was the building of social capital and networks through the formation of voluntary community groups. By ensuring participation and local capacity building, this approach provides the foundation for community-based adaptation and establishes a mechanism through which the community can interact with the outside world. While the community and youth volunteers predominantly worked within the community – forming and reinforcing bonds between community members – the early warning committee CBO provided an opportunity to reach out into existing networks within which development, disaster and climate change information, planning and policy is shared and shaped.

Just as the social networks developed through the project can be seen to have been the catalyst for adaptation activities, so too can the limitations of the project be traced to gaps in the networks to which the community had access. However, underpinning the approach taken by the project organizers was the assessment of vulnerability provided by the baseline studies. The focus on livelihoods and the dependence of the community on natural resources ensured that the project activities remained relevant to the community members. Indeed, the difficulties experienced when trying to disseminate information on the global causes and regional impacts of climate change reinforce the importance of engaging with the immediate needs of poor communities. However, this successful entry point was not entirely capitalized on, as the notion of vulnerability that drove project activities was limited in its scope. While ensuring a focus on appropriate technical activities, failure to engage with the social, cultural and political factors that underpin poverty restricted network building to those opportunities to influence development and disaster planning. Engagement with the broader determinants of vulnerability may have revealed new spaces for network building, such as linking with women's organizations to highlight the gender-differentiated impacts of climate disasters, or seeking partners capable of influencing or delivering resources for scaling up the technologies successfully introduced by the project. Practically, addressing these complex and overlapping factors demands multidisciplinary project teams capable of working together to build a picture of poverty and vulnerability. Moreover, in the *charlands*, access to land following river erosion is a key issue determining livelihood opportunities, and is one that can only be handled at government level.

The focus on local traditional knowledge of weather and climate provided a starting point for introducing the subject of climate change to the community and revealed problems with access to modern sources of weather information. Importantly, this process demonstrated that, while community members had recognized changes in weather patterns, this had not been understood as part of a long-term trend. Similar baseline surveys among government extension workers also revealed a lack of knowledge of climate change. Project activities therefore retained a strong focus on training and awareness raising, ensuring that adaptation work was predicated on a local and institutional understanding of why disaster reduction and livelihood strengthening activities were necessary. However, success in generating a community understanding of the likelihood of long-term climate change impacts also demands that the project puts in place mechanisms for the community to act on this information in the long term. The networks accessed by the CBOs established by the project fulfil this role, offering the opportunity to raise climate change during planning on development and disasters. A significant next step would be to ensure that the community continues to have access to targeted and timely weather and climate change information through the establishment of stakeholder groups where community members and climate professionals could meet, enabling the community's needs to be established and climate information to be delivered in an appropriate and relevant form.

Much of the social network development took the form of building relationships in the context of training and awareness-raising activities. This process repeatedly revealed an important hurdle for community-based adaptation: building communities' knowledge of climate change in order to facilitate and motivate adaptation comes up against the practical problem of engaging poor communities with issues that may not have immediate relevance to well-being. However, climate change training targeted at specific stakeholders (within and outside the community) met with more success, while others in the community or those less receptive to formal training benefit from broad awareness raising. The awareness-raising approach, in which different dissemination techniques are targeted at different groups, illustrates the efficacy of tailoring information and training to the needs and capacities of the recipients. Moreover, the use of local language and the success of cultural shows reinforce the importance of working with existing social norms when introducing new concepts and ideas. The overall success of awareness raising and training will only become clear over time: the readiness of community members to embrace new approaches will be enhanced by their awareness of climate change, but ultimately depend on their assessment of risk. The difference between individuals in this regard is illustrated by the unwillingness of some community members to engage with new technologies until others have proved them to be successful.

Finally, the technologies developed during the project were specifically targeted at the existing problems faced by the community. Implicit in the introduction of innovations such as tube wells, latrines and raised housing is an assessment of high vulnerability and sufficient clarity of future climate change to justify the investment. The vulnerability of the community to future flooding is well established; however, while intense rainfall events are anticipated to continue to increase, as in most regions there is high variation in precipitation predictions for south Asia. In this context it is notable that these are no-regrets technologies in the sense of offering immediate as well as (potential) long-term benefits. Similarly, the livelihood interventions improve diversity of income and therefore resilience, while not locking the communities into a particular development path that may prove unsustainable should climate predictions not play out as anticipated. However, as indicated by the communities themselves, the long-term problems of survival in the ever-changing *charlands* remain – and are likely to become even more challenging in the face of climate change.

CHAPTER 3
Changing seasons and flash flooding in the foothills of the Nepal Himalaya

Prepared with Dinanath Bhandari, Project Manager, Practical Action Nepal

Abstract

This case study discusses a project that sought to address climate change in the Chitwan district of Nepal. The subsistence farming communities in the area have suffered landslides, flash floods, unusual rainfall patterns, seasonal storms and droughts with increasing regularity in recent years, eroding the productivity of the landscape and destroying lives, livelihoods and infrastructure. The project created new community-based organizations that, as officially sanctioned bodies, were able to participate in formal planning and information-sharing institutions, supporting the community's efforts to adapt through the implementation of disaster risk reduction measures. The project gained national attention in Nepal and provided the communities with improved natural resource management skills as well as the confidence to employ a diversity of crops in order to improve the resilience of their livelihoods.

Introduction

Nepal contains five major climatic regions, from subtropical in the south to alpine in the north, split across three main geographical regions: the Himalayan range across the northern borders, a middle hill range and the Terai plain along the southern border. The impacts of climate change differ with altitude. In the high mountains, global temperature rise is leading to glacial retreat, with increased risk of glacial lake outburst floods. Increased run-off variability and sediment loading from glacial melting have an adverse effect on hydropower generation, irrigation and rural livelihoods.

In the middle mountains and lower altitudes the impacts of climate change are less visible, yet they affect a larger population and the ecosystems on which they depend. A growing human population has led to deforestation and the expansion of agriculture to the hill slopes, leading to increased soil erosion and flooding. Degradation of resources, altered land use and changing precipitation have collectively resulted in water-induced disasters. These in turn are anticipated to have further impacts on water availability and quality, productivity and human health (Eriksson, 2006).

Climate change is thus a key element of the overall context of vulnerability that affects every aspect of human livelihoods in Nepal. When land is damaged by flooding, families are forced to cultivate another parcel of land – often clearing forest land either permanently or through shifting cultivation on the hill slopes. This displacement in turn creates additional pressures in the new location, weakening the land's capacity to withstand future climate-related shocks. Climate-related disasters also lead to negative impacts on social systems, forcing changes in occupations and land-use patterns so that entire livelihood options are affected.

The project described here is located in the south-western area of the lesser Himalayan foothills, in a watershed covering over 12 square kilometres. Just over half the area is on fragile soils in the river valley, a landscape based on colluvial deposits, alluvial fans, aprons and ancient river terraces. The rest of the area consists of steep forested slopes. Forest and agricultural lands are the two main uses for the land, which also hosts scattered settlements.

The majority of the population are dependent on the natural resources and monsoon climate for their livelihoods. Agricultural land is mainly of two types: uplands known as *bari*, without surface irrigation, and terraced rice fields known as *khet*, with surface irrigation through channels from rivulets. *Khets* are considered valuable land where farmers can grow at least one rice crop a year. They are generally on the banks of rivers and more vulnerable to flooding and erosion. This area was selected based on the observed changes in weather patterns and weather-related disasters in recent years, reported in consultations with national and district-level stakeholders. The final choice of project site was made based on the fact that the potential impact from weather extremes is more severe than in alternative sites, there is a high level of vulnerability of the communities, and the area is of high priority to the district and local government.

Almost all families in the area were engaged in subsistence agriculture and livestock raising. Only 48 per cent of the total population were solely engaged in agriculture. About 35 per cent of the households posses only uplands (*bari*) where irrigation is not available. Fourteen families had been cultivating lands owned by others on a crop-sharing basis, with strategic management decisions taken by the land owner who generally lives outside the locality. Only 30 per cent of the families in the project area were able to meet their annual food needs from their own production, and more than 40 per cent of the families only produced enough food for three months. Waged labour, working outside the area, and remittances from family members overseas are the main other livelihood strategies. Young people like to go abroad to work and around 100 individuals from the community were working outside the country. A major source of income for many families was liquor production and sale at nearby markets, but this was not reported as it is illegal and not a socially respected activity. Since this activity was not being disclosed, it was not possible to get accurate figures on family income.

Since young men tend to work outside the villages, capacity and responsibility for immediate response to disasters rests mostly on older household heads and women. The majority (61 per cent) of the population are literate, able to read and write in Nepali language. Only 24 individuals have attended university. Most of the illiterate are women. While seasonal sickness is common, outbreaks of diseases are not encountered. Mild malnutrition is common among children from poor family backgrounds.

Group discussions in the area assessed and categorized whole families based on their strengths and resources in terms of assets. The results are categorized in Table 3.1.

Electricity is available in almost all houses. Footpaths link homes to service centres and the main road. During the rainy season mobility is often temporarily limited by flash floods in the streams. Piped drinking water is available through community taps for more that 80 per cent of the population, although intakes are vulnerable to flood and landslides; irrigation channels are similarly affected.

There is a well-stocked natural forest dominated by *Shorea robusta* (Gartn). People collect fuel wood, foliage for fodder and bedding for livestock, and timber for the construction of houses and cattle sheds. Some groups, particularly the *Chepangs,* largely depend upon the forests for food and income. They collect wild fruits, twigs, tubers and plants for their daily use, and fuel wood for

Table 3.1 Wealth ranking

Wealth ranking	Proportion of households (%)	Indicators
Well-off (A)	31	Enough cultivated land (*khet* and *bari*) Food sufficiency for the year Income from service/pension/business/remittance Small family size Employed Large number of livestock
Medium (B)	36	Smaller cultivated land (only *bari*)/low production Food sufficiency for only 6 months Borrowing from village money lenders Lack of sufficient income sources Partial dependence on daily wage labour Larger family size Seasonal unemployment
Poor (C)	33	Smaller cultivated land (only *bari*) Very low production/food sufficiency for 3–4 months a year Unemployment Lack of skills (capacity) for alternative income Larger family size Heavily depend on daily wage labour

both domestic use and sale to buy groceries and clothing. The environmental services provided by the forest are perceived to be important.

Community-based adaptation in the middle hill region of Nepal (Chitwan District)

Project activities were initiated from October 2004 and ended in December 2007, focusing on the area managed by the Kabilash Village Development Committee (VDC hereafter). Project activities covered about 200 households of mixed communities scattered in 10 different settlements. Awareness raising and advocacy campaigns targeted the VDC, district and national-level policy makers, development agencies and media. Further awareness activities on climate change and associated issues were carried out through teachers and students in 12 different schools in the district.

The project was implemented with an ad hoc committee of the communities which was developed and institutionalized into a community-based organization (CBO) in the second year of the project. Ecological Services Centre (ECOSCENTRE), a local NGO, provided agricultural inputs to farmers with financial support from the project. For awareness raising and policy advocacy, the project worked with different government administrative units including the VDC and NGOs at the district level, and the Ministry of Environment, Science and Technology and different networks of NGOs at the central level.

Climate change in the area

The project area has a subtropical climate. Data from the nearest two meteorological stations (about 20 km south-east and south-west, in the same climatic zone) reveal that average precipitation has increased by 614 mm and average temperature has increased by 1.3°C over the last 30 years. More than 80 per cent of the total precipitation is received in the three months between the second week of June and first week of September. When asked about their perceptions of climate changes, 98 per cent of respondents noted changing climatic conditions in the area, particularly increasing drought and an unusual rainfall pattern (95 per cent of the respondents) and temperature rise (5 per cent of the respondents). Local people also reported that rainstorms have decreased in number but increased in intensity.

A process of family interviews, successive group discussions and observations identified landslides, flash floods, unusual rainfall patterns, seasonal storms (dry winds, hail and thunder) and droughts as major hazards increasing in recent years. For the past few years, people have observed long lasting winter fog (cold wave) which had not previously been seen. Higher temperature in the daytime, particularly during summer, affects work in the fields. The frequency and magnitude of the disasters has increased over the past decades as Table 3.2 shows.

Table 3.2 Flood disasters in the project area (from field survey and VDC records)

Event	Impact
1966	People and livestock killed, land and houses damaged (actual figures not available)
1983 July	About 2 hectare khet and a grinding water turbine were damaged
1993 August	About 5 hectare khet of three families was washed away and three people killed
2003 August	Ten people, more than 80 goats, 5 buffaloes and 3 oxen killed. Five homes and five cattle sheds destroyed. Over 12 hectare khet washed away. Drinking water supply system to the district headquarter city completely destroyed
2006 September	Four houses, four cattle sheds, one footbridge, one village drinking water supply system, seven irrigation channels destroyed and over 20 ha khet damaged

These hazards have significant impacts on the livelihoods of the communities and the ecosystem. Erratic rainfall has multiple adverse impacts. Heavier rainfall over a shorter period prevents the proper recharge of watersheds, and the available precipitation is lost through run-off and overland flow. This in turn creates landslides and flash floods, erodes fertile soil thereby decreasing productivity, and ultimately degrades or destroys livelihood assets. This erratic rainfall creates seasonal scarcity of water for irrigation during the non-monsoon season, while long gaps between two successive rains (even during the rainy season) can lead to increased frequency of droughts. Moreover, erratic rainfall prevents timely planting, care and harvesting of crops and thus decreases food production. Traditionally farmers have maintained a calendar of cropping and harvesting based on the rains and seasons, but now this calendar does not fit recent weather patterns.

Dry wind storms used to come during the pre-monsoon season between March and June, but people reported that they can now occur at any time of the year. Dry storms often destroy the roofs of houses and cattle sheds, damage crops and injure people. Hailstorms have also become common but unpredictable events in recent years. They destroy crops and corrugated zinc roofs, kill wild birds and destroy their nests and eggs, which results in an increase in insect pests in crops. The trend to longer winter fog in the valleys causes wilt in winter crops such as potato and mustard oil.

When a community asset such as a footbridge is destroyed it has a widespread impact. For example, a flash flood washed away a footbridge in 2006. This was the link to school, local and district-level government offices, the market and the health centre for both the upstream and the downstream communities.

Social networks

People from 12 different ethnic and caste groups live harmoniously in the project area. Communities are tied by social relationships that cross caste and ethnic groupings. They are organized into informal groups such as mothers' clubs, youth clubs and cultural clubs. However, these groups do not have any capacity to help cope with disasters as they lack awareness, financial resources and relevant skills. There are two primary schools and one secondary school in the area, with local government offices and other service providers nearby. However, health services, agricultural extension advice and inputs suppliers are inadequate.

The project helped the informal local groups to become part of a broader based organization serving the whole community. Training was given to the members of the community-based organization (CBO) on book-keeping and organizational management. Regular mentoring and monitoring was carried out on rules, procedures and responsibility for procurement, decision making, social audit and so on. A constitution was prepared in consultation with each group and a final draft was presented for endorsement. After the group passed their constitution, thereby establishing a community-based organization, the executive committee of the new CBO applied to the district administration office for registration in line with the prevailing government law. The group received its registration in the second year of the project. This provided the community with a legal basis for receiving funds and planning and implementing activities as an independent body. Linkages and partnerships with local government, NGOs and networks at the local and district level were later developed, which provided the CBO with an opportunity to share experiences and gain access to information. A further component of the local social network created by the project was the creation of three forest user groups, considered in more detail below.

Awareness raising

Raising awareness on climate change was considered an important aspect of project implementation. Information on the impacts of climate change was written up to fit the local context, and a booklet was published in the Nepali language and disseminated among teachers, students and community groups. Presentations were given in classrooms for older secondary school pupils and members of informal groups. A poster was published and disseminated widely. Further awareness activities on climate change and associated issues were carried out through school teachers and students in 12 different schools in the district, reaching over 2,700 students. Twenty-five schoolteachers were trained for three days and assisted with resource materials before going on to teach students of grade eight and nine (between 14 and 17 years old) in their schools through extra-curricular activities and excursions once a week for six months. This initiative was widely appreciated and national television broadcasted a news feature of this work as an example of good practice. Other organizations have also replicated this model in different ways.

At district level and beyond, five workshops were organized in different cities in Nepal. Information on the impacts of climate change and importance of community-based adaptation was shared among policy makers and civil society organizations. Project staff also worked closely with the Ministry of Environment, Science and Technology, the focal ministry for the UNFCCC, for awareness raising and updating on recent issues in international negotiations. As a result of these efforts, the Ministry has selected Practical Action Nepal as a member of its recently established national climate change network.

Project staff participated in network meetings and workshops to share project lessons at both national and international level. This was deemed one of the most effective project activities in terms of communicating climate change and adaptation information. As a result ECOSCENTRE has revised its programme objectives to include climate change as one of its areas of work. Staff in the organization are now acting as local resource persons for sharing climate change issues among stakeholders. Following completion of the project, most NGOs in the district have broadened their activities to establish linkages with climate change issues, particularly on adaptation and disaster preparedness. NGOs, Rotary clubs and government line agencies have started to organize workshops and sharing events; Practical Action is often invited to give presentations at these events. Journalists have also published articles and case studies in newspapers as a result of meetings with project staff, and the project work has also featured in national and international magazines.

Participatory technology development

The project process initially involved discussions to identify local problems, needs and possible solutions. Participants were invited to prioritize activities, and those to be implemented in the coming year were identified. The group meetings also revealed activities beyond the scope of the project, including: construction of school buildings, road and bridge construction, and support for additional teachers in schools. A brainstorming session was used to identify potential sponsors for these activities and suggestions were given to the community on the appropriate agencies to approach. As a result of this process, funding and construction of additional school buildings was achieved. The implementation of each activity on the priority list was discussed individually at the community level, with the result that costs were estimated and an annual work plan prepared. The aim was to foster community ownership of the plan.

The baseline survey and subsequent participatory discussions revealed the need for improvement in existing agricultural practices which would require skills and input support. Project training focused on three areas of agricultural technology: vegetable growing, livestock raising and soil, forest and watershed management.

Table 3.3 Training events

Training activity	Number trained	Male	Female
Vegetable growing	50	36	14
Exposure visits to other villages	25	13	12
Exposure visit for Chepangs	23	16	7
Forest management	45	28	17
Slope stabilization and cultivation	30	18	12
Goat keeping	25	20	5
Animal health training	3	3	0
Regular support on farming techniques (farmers field school)	20	14	6
Organic farming training	11	3	8

Three youths participated in 5-week basic livestock raising training to become village animal health workers. One now runs an agro-vet shop and provides technical services to farmers such as vaccinations and advice on fertilizer and pesticide application. Two individuals participated in meteorological station management training. One community member attended a 2-week long organic agriculture training course in order to act as a demonstration farmer for others.

The project approach focused on increasing expertise to manage local resources to build resilience, and on diversification of crops. There is a double benefit to diversifying crops: firstly, it provides farmers with multiple opportunities for food production and income generation; secondly, increasing agricultural biodiversity coupled with improved soil and water management provides other ecological benefits to the system. With improved absorptive capacity of the soil, the natural system is more resilient to heavy downpours, preventing soil erosion and maintaining watershed health.

One of the problems in the project area was reduction in the area of productive land because of flooding and lowered production as a result of erratic rainfall. The majority of the farmers were unwilling to change their traditional monsoon cropping pattern of rice and/or maize, despite falling yields. However, farmers showed interest in growing short rotational crops, particularly vegetables, in the winter when fields were generally left fallow. This was a viable option as there was an accessible market close by. Techniques for growing vegetables, managing sloping land and approaches to efficient water management were all new to the community. The technologies were introduced as demonstrations and then expanded to some selected farmers; following this, they were widely replicated among households. A second problem was that the changing pattern of rainfall prevented farmers from obtaining successful harvests from rice growing, and so some farmers were looking for an alternative way to augment incomes. For this group, the option of introducing fruit trees was put forward.

The following technological developments were introduced during the project:

Vegetable production

Training was provided for 25 farmers representing each settlement. Following the training farmers were supplied with vegetable seeds obtained from the local market and demonstration vegetable plots were established. In the first year a few innovative farmers initiated cultivation, and the project provided technical support by paying the costs of agriculture experts to visit the area frequently for one year. The extra income these farmers earned, which compensated for lower cereal production, encouraged many more farmers in the area to come forward in the second year and the area of vegetable cultivation increased markedly. In the final year of the project almost every farmer took up vegetable gardens. Small producers used the crops for household consumption, while those with larger plots were able to grow for both domestic consumption and sale. For both groups, vegetable production has saved money usually spent on vegetables and led to a better diet and thereby to improved health.

Fruit production

Field observation visits to banana and fruit orchards were arranged. The project purchased 2,000 fruit tree seedlings from nearby nurseries, and supported farmers with advice on planting and care of the trees and assisted with marketing of the products. Two farmers have switched to banana and fruit cultivation as rice was failing to produce sufficient yields owing to the lack of irrigation water. Many other fruit seedlings of mango, lychee, citruses and coconut were distributed in the village and were at the nursery stage by the end of the project.

Irrigation channel improvements

The intakes of irrigation channels were often damaged by flood and landslides during the summer, while during the dry season streams almost ran dry. The small amounts of water available at the intakes did not reach fields because of leakage in the course of flow. To overcome these problems, farmers and project staff identified concreting of channels and channelling through pipes in some difficult points as appropriate solutions. The project provided materials (cement and high density polyethylene pipe) and skilled labour, while the community contributed labour and local materials (sand and stones) for these infrastructures. Eight small irrigation channels were repaired or improved and their intakes were strengthened with gabion walls. This work provided improved access to water during the summer for rice cultivation and during winter for vegetables, bringing benefits to 214 people from 42 households, enabling them to irrigate 9.6 ha of *khet*.

Goat raising

The project provided training on goat raising to 25 farmers and provided one female goat for breeding to each participant. The CBO set up a mechanism by which each recipient provided one female to her or his neighbour after the first generation of kids have matured. Three breeding bucks were also provided, one to each goat-breeding group, to overcome the problem of inbreeding and improve characteristics such as faster growth and larger size. Each participating member paid 100 Nepali rupees (US$1.28; exchange rate NPR1 = $0.013, 9 December 2008) to the CBO once a kid had been born. This amount is utilized as insurance against the death of a participating member's goat (for reasons other than negligence). At the time of writing five farmers had benefited from this community insurance scheme and the exchange programme had reached 62 families. Offspring from the improved breeding process have been produced and have generated income after being sold for meat.

Check dams and flood barriers

Landslides and floods sever lands and cover soil with debris. At the beginning of the project, local people proposed building dykes and dams downstream. However, the project provided information and organized exposure visits to areas where people had conserved the watershed through integrated approaches. These visits raised awareness of the benefits of blocking potential torrents at source (while the flow remains small) and the community agreed to work at upstream locations. After observation of the results of good practices, people in the project area were convinced to work in the micro catchments which feed to the main stream. Gabion check dams and retaining walls were constructed in the tributaries and dykes in the main stream. Planting was carried out in the erosion prone areas. A severe flood in 2006 caused some of the dykes in the major streams to be washed away or broken down, but the structures in the tributaries were able to stem the flood and provide stability. Local people, with the material and technical support from the project, continued constructing check dams which have enabled them to reduce the risk of disaster due to flood.

Outcomes of the training and technologies

By the end of the project, vegetable production had overtaken liquor as the largest source of income. Water resources were better utilized to support livelihoods: land that had remained fallow during winter was cultivated to grow vegetables and, where appropriate, some farmers switched to growing banana. Farmers have been able to adapt to new varieties and modes of irrigation and production in different seasons. They have made choices based upon their circumstances. Previously, land used to remain fallow if rice could not be transplanted during the summer. Improved irrigation facilities, and the use of water saving technologies now enable the planting of other crops in the next season. The most important outcome is that people have the confidence to make different farming decisions.

By the end of the project farmers of all levels had improved food security and increased resilience to cope with adversity compared with the baseline. However, households with medium well-being had improved the most. Overall, household food security has been increased by between one and six months through increased production.

One of the trainees opened an agro-vet shop and (in 2008) was providing primary health care services to the livestock in the area. He supplies seed for cereal and vegetable crops and technical advice to the farmers, offering a cheaper and more reliable service than was previously available. Twenty-one farmers (out of 30 trained) have practised sloping agriculture land technology on their shifting cultivation patches while five farmers stopped the practice and have planted fodder on the slopes. Eighty farmers benefited directly from training and the same number indirectly through replication and exchange of services and information.

Forest conservation

In recent years, the forest resource has been degraded for various reasons including illegal logging. The problem was highlighted during the baseline survey. The project supported the communities to form and strengthen community forest user groups. The community began patrolling the forests, raising awareness and curbing illegal logging in cooperation with the district forest office. These initiatives improved the condition of the forest, which encouraged the community to think of a long-term strategy. The three forest user groups prepared their constitutions in line with the forest law and regulations and registered their groups with the district forest office. This provides a legal basis for each group to conserve the forest area as defined in the constitution. For this, an operational plan is required and the communities have begun to prepare a strategy with the support of district forest office staff. When the plan is approved the groups will have the right to manage their forest.

A key outcome of improved forest management is the restriction of illegal timber smuggling by outsiders. Collection of firewood has been regulated. Only local people who depend on forest resources, such as the *Chepangs*, are allowed to collect and sell dried wood. Timber can be made available for construction of social infrastructure such as footbridges and school buildings, but local people who want timber are subject to a needs assessment by the community and pay a reasonable price for it. One of the three forest management groups is a team of women, providing an opportunity for women to develop leadership in local natural resource management. With the built-in linkage with the district forest office, representatives of each group receive opportunities to participate in training, workshops and exposure visits. Local government also recognizes the groups, which provides further encouragement to better manage the resources.

Resource implications

The project included two full-time staff from Practical Action, a project officer (in a management role) and a community mobilizer, and a team leader based in Kathmandu worked part-time on the project. The project employed a district-based NGO (ECOSCENTRE) to provide services to the communities on conservation-friendly agriculture, vegetable growing, cash crops and livestock raising. For this work, two people were partially engaged in the project for two years. In addition, 15 individuals from outside the project area were involved in providing different training and capacity building initiatives to the communities. These individuals were involved mainly in skill improvement training and came from government offices, NGOs working in the field of agriculture and independent consultants. In terms of material inputs, the project initially provided seeds and agricultural implements, but in the second year, many farmers bought their own seeds in addition to project support. For construction works the project provided skilled labour and support for purchasing materials and tools which would not be available locally.

The total budget was £67,600 ($99,700), and there were 1,124 beneficiaries. Materials accounted for £11,900, about one-sixth of the total. The cost per beneficiary was £60.

Lessons and challenges

Building the confidence of farmers is critical. The most important factor in achieving this was showing farmers that a new approach can work better than the traditional techniques in changing environmental conditions. For some participants, the level of training was insufficient to ensure successful adoption of the technologies: for example, fruit growing requires several seasons and constant follow up to become profitable. Owing to a lack of land, some farmers were not able to establish commercial farms and wanted to grow a few plants in the terraces or in front of their houses to increase income. Similarly some of the participants were not able to influence other members in the family to implement what they had learned in the training. Such families waited to see whether their neighbours obtained success with the intervention. In contrast, though, one or two farmers learnt the technologies so well that they have become successful fruit producers and sellers, working their way out of dire poverty. Demonstration helps to foster replication. People generally learnt from observation and sharing experiences with their neighbours. When they are able to see the process and results they are more confident of the effectiveness of the initiative than when based on classroom learning or theoretical training alone.

One really effective achievement was enabling a CBO (the Climate Change Impacts and Disaster Management Group) to become a formally constituted and registered organization, with officers trained in book-keeping. This has enabled it to coordinate activities implemented by other agencies in the area and to build partnerships with village level government in implementing annual activities. It succeeded in giving small loans to some farmers for vegetable growing out of its own funds raised from the goat insurance scheme and income from breeding services. The CBO has been included as a member of the district disaster mitigation network and one member from the CBO is a representative at the executive committee of this network. This provides opportunities to share experiences and learn from others on disaster prevention and mitigation related to the extreme climate events in the area. Linkages have also been established between the local government (VDC) and government service providers including district agriculture office, district livestock service office, district forest office and other sectoral offices and NGOs. Influenced by project activities, the VDC council has allocated 10 per cent of its budget to watershed conservation.

Besides the formally constituted CBO and the forest user groups, there were several other informal groups established such as goat-keeping groups, irrigation management groups, and vegetable growing groups. These informal groups provided a space to share and learn for individual farmers.

Selection of appropriate seeds and varieties determines the success of harvests. It was a challenge for farmers to identify the best variety and quality seeds in the markets. Close linkage to service providers is therefore important. When farmers and seed sellers have close linkages and farmers are able to influence the trader, the latter becomes responsible for the seeds they sell to the farmers and risk of cheating is reduced. For this reason, three local youths were trained in agro-vet skills. If the seeds supplied by the local supplier do not do well, the local people have the skills to criticize. As the agro-vet's enterprise is dependent on the quality of seeds, a powerful incentive to provide a good service is generated.

One significant challenge of implementation was the varying levels of participation in different activities. Community participation was higher in the activities which directly benefited individuals and which had clear boundaries as to who would benefit, such as repair of the intakes of irrigation channels. Participation was very low or needed very intensive mobilization for the activities with wider or less direct benefits, such as construction of check dams and repair of footbridges. In some communities where there is less of a tradition of working communally, people need to be motivated to understand the benefits to all of, for example, a functioning community infrastructure. In such circumstances, awareness raising on the joint responsibility to manage resources wisely for their sustainability may be appropriate. This takes time (in some cases beyond the project period); however, working in groups is the only viable approach to organize people to achieve common goals.

Certain outcomes will be key for the sustainability of adaptation. While good informal linkages have been established between the community and different service providers, such linkages need to be maintained, and the dynamic nature of institutional change in Nepal makes this a challenge. The role of the local CBO is crucial to help communities in this regard.

There was a lack of experience among farmers on seed selection and production. While links to the local agriculture office can help with advice on new varieties, maintaining their own seed supply gives farmers a degree of independence from inputs markets. This activity was not possible during the project period as farmers were not confident and interested enough to become seed producers. Government and other service providers need to be influenced to identify and adopt sustainable options for seed availability.

Conclusion

In this project, the community was motivated to take part because of the increasing frequency of weather-related extreme events that had been affecting their livelihoods. Awareness raising on climate change, and making the links with disasters and local disaster risk reduction options (that would also help with adaptation) was key to involvement. Understanding the linkages between socio-economic activities and their impacts on the environment was crucial for changing behaviour towards managing local resources in a sustainable manner. For this reason, there is a need to raise the level of awareness on climate change and its impacts among all sections of the community.

Strengthening networking capital by forming legally constituted organizations for disaster preparedness and forest management, and linking these into the formal government institutional framework was vital for ensuring the future viability of activities, and depended on raising awareness on climate change issues within the local government system as well as within the community.

Adaptation to climate change requires a larger and longer-term investment in capacity building than the three-year project approach allowed. Where the environment has become degraded through the long-term use of unsustainable activities, such as slope cultivation and shifting cultivation in forested areas, sufficient time to develop alternative strategies is needed. In this project, improved natural resource management was required, as well as farm-level changes to production systems. An integrated approach for development and ecosystem management is essential while devising and implementing adaptation activities. Cultural and social issues around resource use need to be understood before changes can be introduced. For example, the *Chepangs*, traditionally a forest dwelling people, were given more rights to collect wood than other groups. In some instances, major initiatives may be needed but elsewhere even minor changes to current patterns of resource use and management (such as use of land for winter vegetables) can provide a 'no-regrets' option to strengthen livelihoods and increase resilience to climate change. This led to a high level of take-up – though only after more cautious farmers had observed the success of their neighbours, highlighting the importance of demonstration. However, there is a risk that people, having found an income-enhancing strategy that works for them now, will not actually have appreciated that adaptation requires a continual process of reviewing livelihood strategies and environmental changes and experimenting on a regular basis with new crops.

Working together within the whole watershed has brought home to these communities the close relation between upstream and downstream activities, and how these can have an impact on the livelihood assets of other people. Community or individual farmers experience the effects of climate change without being aware of the local consequences because of a lack of communication channels informing them about what has happened downstream or in an adjacent river catchment. Understanding linkages between socio-economic activities and their impacts on environmental factors is important for the management of local resources in a sustainable manner.

The interlinkages between local topographical factors – weak geological structures, steep slopes – and socio-economic activities such as deforestation, faulty agricultural practices and weak institutional arrangements for land management mean it is impossible to isolate adaptation activities from the need for improved natural resource management and disaster prevention work.

CHAPTER 4
Desert and floodplain adaptation in Pakistan

Prepared with Abdul Shakoor Sindhu, Principal Coordinator, Rural Development Policy Institute (RDPI), Islamabad, Pakistan

Abstract

Climate change combines with failed agricultural policies, marginalized communities and a harsh environment in rural Pakistan. This chapter reviews a project that worked with communities in two locations in Punjab, one in the Thal desert and another on the eroding banks of the Chenab River. Activities sought to work with local culture to enable community organization and awareness raising, including using traditional forums and establishing resource centres and farmers' festivals to help build relationships between different stakeholders and facilitate the exchange of knowledge and views on meeting the challenges of climate change. Alternative livelihoods and agricultural practices were successfully introduced and in the process lessons were learnt as project team members and communities alike sought to understand the complexities of adaptation interventions.

Introduction

This project worked with poor communities who had been identified as vulnerable to existing and future climate change in two very different ecological zones of Punjab, Pakistan. Two of the communities were in former rangelands in the Thal desert region. Before the green revolution in the 1960s the area had been managed by the local communities as communal grazing land. The major livelihood activities were camel, sheep and goat rearing, with small areas of crop cultivation largely dependent on local wells (known as *Khoo*). However, grazing rights were severely restricted following a declaration by the provincial government that designated the community grazing lands as a reserved or protected forest area. Following the publicly funded construction of irrigation canals, government policy was to promote arable farming, and local agricultural extension workers encouraged the former pastoralists to switch to farming, using hybrid seeds and chemical fertilizers in the newly irrigated areas. In areas not served by the canals, people started clearing the natural vegetation to make way for the cultivation of gram on large tracts of land. Where possible, ground water was accessed by means of diesel powered tube wells to start the cultivation of wheat, barley, cotton and fodder crops, yet gram remained the central crop. The non-canal-irrigated areas of Thal have subsequently become one of the biggest gram producers in Pakistan. However, a good harvest of gram depends entirely on timely and sufficient rains. As rainfall is increasingly uneven and unpredictable, many farmers now consider gram cultivation a risky undertaking. Moreover, the clearing of natural vegetation in the quest for large-scale agriculture has proved detrimental to the sustainability of local livelihoods and ecology. In recent years, as a result of irrigation, ground water has been much reduced, while the use of pesticides, chemical fertilizers and diesel has made the communities dependent on increasingly expensive external inputs. In the five or six years before the project began, the area experienced a severe drought. Crops failed, and both livestock and arable systems suffered: natural vegetation had already been cleared, so there was no grazing land, and the bare land became increasingly prone to wind erosion that degraded soils and reduced yields.

The second project site was in Bela, on the banks and in the floodplain of the Chenab River. In this region, the river has eroded productive lands and occasional flooding has displaced communities. This has generated different impacts on different communities in the Bela region: in Yake Wala village, one of the project sites, around three-quarters of the land available for farming and housing has been lost to erosion, while in another nearby village a change in the course of the river has forced the entire community to move from one side of the water to the other. While the geographical location of the communities renders them vulnerable to the effects of flooding, there is also a history of institutional neglect in the area. Areas around the river have been given the lowest priority for development, as agencies have judged investment in these areas too risky owing to the threat to land and infrastructure from flooding. The underdevelopment of these communities had left them marginalized, compounding their vulnerability. Moreover, the isolation of the communities made them targets for groups that have prospered in remote and wilderness areas by stealing cattle. The accumulation of these overlapping problems has left farmers reluctant to invest their labour and capital in their land. Only a few crops are grown in the Bela region and the poorest prefer not to keep livestock due to the risk of cattle being stolen.

As the following sections describe, the project commenced with comprehensive research into local vulnerability to climate change and focused on raising awareness and building social capital to enable the communities to respond to the challenges. Technical interventions focused on livelihoods, aiming to build resilience and provide alternative sources of income. The project worked in four village locations: in the Bela region Bahar Wala village (in District Sargodha) and Yake Wala village (District Jhang); and in the Thal region in Makhni Khoo and Wasawa Khoo villages (District Layyah). In total the project reached around 300 households. The project was undertaken by a team from the Rural Development Policy Institute (RDPI), an Islamabad-based non-governmental organization, with technical and financial support from Practical Action and the Allachy Trust UK, respectively.

Community-based adaptation in rural Pakistan

The greater portion of the first year of the project was dedicated to identifying the beneficiary communities within the localities, understanding their situation and building a relationship with them. Baseline information was collected using techniques including questionnaires, focus group discussions and in-depth interviews, and served both to provide information to the project staff and build links and trust with the communities. The baseline information was sufficient to contribute to project design, but the field teams noted that they were unable to address some issues as a result of deficiencies in their understanding of climate change. Climate change and research training sessions had been arranged for the field teams, but it was only as the project progressed that the teams became sufficiently confident in their understanding to be able to fully engage with the communities during the research process. Progress was inhibited by the lack of climate change and adaptation literature in local languages: the majority is in English or other international languages, which the field teams were insufficiently familiar with. Their understanding of the issue was built largely through conversations with the project manager and with the subsequent production of material in Urdu, Punjabi and Seraiki.

Assessing climate knowledge and awareness

Gathering local knowledge on how local people perceive and experience climate change formed the first step in involving communities in the adaptation process. Local knowledge was collected primarily from elderly men and women of more than 60 years of age through in-depth interviews with respondents. Rather than immediately asking about changes to weather patterns, the interviewers first tried to establish the characteristics of weather in the area. Only then were the respondents asked about the changes they had witnessed in current weather patterns compared with those that they had been used to in earlier decades. Questions were framed in light of previous field experiences and a review of literature on climate impacts in Pakistan. The questionnaire touched on broad social issues as well as the changes that had occurred in weather conditions and the impacts on people's livelihoods and living patterns. This approach helped the research team generate a rapport with the respondents, and both sides found the informal conversations interesting and informative.

During the collection and analysis of information it was found that communities were (understandably) unaware of the scientific dimensions of climate change. However, almost all the respondents were of the view that the local climate had changed and that this had led to changes in their livelihoods. In particular:

- There had been a notable increase in the intensity and length of the summer season.
- There had been a decrease in the intensity and length of the winter season.
- Some 25 to 30 years previously there had been four seasons: winter, spring, summer, and autumn. At the time of the project, the respondents felt that there were now only two seasons – summer and winter – with summer lasting for 9 months and winter reduced to 3 months.
- Uncertainty about the rains had increased. It had become impossible to say when it would rain and when it would not.
- A number of flora and fauna species had become extinct during the previous few decades.
- Local knowledge about weather patterns and early warnings had been rapidly eroding as the new generation was not interested in learning the techniques. Many signs and warnings that had previously proved effective had become unreliable in recent years.

Social networks

Community mobilization was a continuous activity throughout the life of the project. Community-based organizations were developed with a view to giving them the lead in identifying, designing and managing the project activities. It was also intended that the community organizations would be transformed into Citizen Community Boards (CCBs), enabling them to gain registered status and be eligible to receive funding from their respective local governments for development work in their areas. It was envisaged that this transformation would help ensure the smooth exit of RDPI from the communities. However, while community organizations were formed at all the project locations, they were unable to gain CCB status until very late in the project, owing principally to the slow response of the local government and widespread corruption in the registration authorities. Further, the community organizations were neither enthusiastic about nor capable of meeting a requirement for CCBs to contribute 20 per cent funding for new project ideas, with the remainder met by the local government. While it was suggested that RDPI could invest the required 20 per cent out of the project budget, the idea was resisted by the project team on the basis that it would kill the spirit of community participation. The prevailing view was that communities would not own the interventions unless they contributed towards their own well-being.

Despite these problems the community organizations at both Yake Wala and Bahar Wala appeared very effective in undertaking project activities. While the community groups at Thal failed to meet expectations, alternative methods of community mobilization were more successful. For example, the introduction of people's assemblies or courts (*Lok Sath*, more on which below) was effective in Thal and attracted the participation of both men and women. The field team in Thal also introduced activities including story nights, camel carnivals and fruit collection parties, which proved effective in bringing community members together. Analysis by the project team revealed that the success of these events compared with the introduction of CBOs was due to their cultural significance: they were activities embedded in centuries of tradition.

It is notable that one activity that was not planned as part of the adaptation project sparked enthusiasm among the communities for collective action. During the baseline survey, communities at both Bahar Wala and Thal ranked support for education as a high priority. There was no school In Bahar Wala and many school age children were either not attending school or travelling 4 to 5 kilometres to reach schools in neighbouring villages. In the project locations in Thal, two young people were running community-based schools and requested support. Mindful that this was a demand that had come from within the community, RDPI decided to help people establish a school at Bahar Wala and provide support at Thal. This activity had the following benefits:

- The community donated land for the school at Bahar Wala and thus showed a commitment for undertaking joint activities. This process helped organize the community into a CBO.
- A local youth at Bahar Wala – one of the very few educated community members – was prepared to serve as a teacher and subsequently became a volunteer social organizer for the project.
- The schoolteachers at both locations in Thal also proved to be the most effective social organizers for the project and were among the first to demonstrate the technologies promoted during the project.
- The schools provided the project with a mechanism for educating children on the environment and climate change.

Awareness raising

The project devised and experimented with a number of awareness-raising tools. The target audience included the communities, local government, civil society and provincial and national government departments. The tools achieved varying degrees of success in attempting to overcome a number of significant challenges:

- There was very low literacy among the communities in general and among women in particular.
- Civil society was indifferent to climate change at the start of the project.
- Local governments had very little understanding of climate change and disaster risk reduction.
- A large difference was noted between the experiences and expectations of the community and the priorities of policy makers. For example, the

government has plans to construct a multi-billion rupee greater Thal Canal to bring more land under cultivation. However, there is significant resistance to this project from the local population as they fear problems such as the acquisition of their lands and the arrival of settlers with associated social conflicts.
- Few resources were available for awareness-raising activities.

Within this context, the project employed a variety of techniques to raise awareness among different stakeholders. Many of these activities were opportunities for people to meet and discuss issues and were thus frequently also mechanisms for building social networks.

Baseline survey as a tool to raise consciousness and awareness

The research questions were used to create a debate and raise awareness. For example, the focus group discussions evolved into a debate and two-way learning process in which the project team better understood the needs of the communities, and the communities developed an awareness of climate change. Similarly, the interviews with local government officials helped raise their awareness of climate change.

People's Parliament (Lok Sath)

The *Lok Sath* is a traditional institution through which communities collectively resolve disputes. The project team in Thal found that the institution still existed but its importance had reduced over recent years. After consultation with the communities, it was decided to revive the institution so that it could be used as an open platform for communities and other stakeholders to resolve differences. For example, in Thal the target communities repeatedly complained that their traditional common grazing lands no longer belong to them. They revealed that around 30 to 40 years previously the grazing lands were handed over to the forest department, restricting communal grazing rights. The *Lok Sath* was used to bring the communities and government departments together to discuss this issue. While the communities were willing to participate, the government officials sought to avoid being part of the forum, apparently fearing public accountability. However, the *Lok Sath* was successful in raising awareness of climate change, promoting public accountability, and providing a socially and culturally acceptable discussion platform. Moreover, the *Lok Sath* helped build the knowledge of the project team as issues were raised in the forum that were not discussed elsewhere.

Story nights (Kahani Raat)

Story nights or *Kahani Raat* are a centuries-old instrument of communication and entertainment in the Indian Subcontinent and Middle East. Before the widespread availability of mass communication, storytellers were in high demand. The project team in Thal decided to revive this tradition and use it as a tool for education and awareness. For example, one such story night was organized on the eve of World Biodiversity Day. Two storytellers related stories in which all the characters were animals and plants. This was very popular among the desert population, with both women and men participating. While the events were successful in Thal, the project team in Bela resisted the idea owing to the danger of criminal activity during the night-time.

Discussions and community meetings

Opportunities for low-key communication and one-to-one meetings proved effective for all stakeholders. Open discussions and informal community meetings were a feature throughout the project, and were cost-effective, purposeful and successful. The approach reflects oral traditions that are still strong throughout Pakistan, with many preferring face-to-face interaction and communication.

Cultural events

One major contribution of the project was the revival and support of local environmental festivals. In Thal, *Salvadora persica*, or *jaal*, is a native tree species that is highly adapted to desert ecology. It has fruit that can be eaten by humans and leaves that are fodder for livestock. The wood of the tree is traditionally used as a toothbrush (or *miswaak*) and is also used for making highly valued decorative items. However, despite its usefulness this native tree is facing extinction in Thal. RDPI decided to make people aware of its importance by reviving a local celebration in which groups of people would go into the wild to collect *jaal*'s fruit (*peelu*). Similarly, camels are desert animals that are highly resistant to dry conditions. In Thal, the camel has a significant role in local people's livelihoods, food security and transport. Its wool has been used in making rugs and ropes while its skin has been used in making expensive decorative items. However, over time the habitat and vegetation on which camels depend have been eroded. RDPI decided to raise awareness among the local population in Thal about the dangers posed to the camel population by organizing traditional camel dancing events and a side event at a local camel carnival (one of the biggest in the Indian Subcontinent). These activities attracted a lot of attention and proved very effective in raising awareness of the value of indigenous species.

Celebrating international days

The project has pioneered the celebration of international days at the local level. In Pakistan, the assumption is that these occasions are celebrated by organizing a conference or seminars at expensive venues. The project tried to break this trend by taking the international debates and issues to the people. International Rivers Day, World Disaster Reduction Day, Women's Day, International Environment Day and World Biodiversity Day were all marked locally. For example, the venues for the International Rivers Day were chosen to be the banks of the River Chenab (in Bela) and the River Indus (in Thal). The venue for International Biodiversity Day was a desert location in Thal, the International Environment Day was celebrated in local schools, and Women's Day was celebrated in women's colleges in both Thal and Bela. All these events were celebrated with minimal possible resources, with RDPI providing only a modest meal.

Local language banners and publications

Literate community members and stakeholders were targeted with materials such as posters, brochures and newsletters. Three posters were published by the project and generated significant interest. The project's flagship publication, *Rut* – which covers climate change and environment issues in Urdu, Seraiki and Punjabi – generated a wide readership. It represents the first publication in Pakistan to highlight local climate change issues and adopts a lively format using stories, poetry and picture stories. The project also produced pioneering publications including a local knowledge report, documentation of the flora and fauna of Thal and translations of introductory information on climate change.

Documentaries

A video documentary detailing the project activities and climate change in Pakistan was prepared during the project. The programme was screened at a number of events, including the National Consultation on Climate Change (organized by RDPI and the Ministry of Environment) and District Policy Dialogues on Climate Change and Disaster Risk Reduction (organized at Hafizabad, Sargodha and Jhang). The documentary has also been distributed to stakeholders and is the only product of its type available in Pakistan.

Campaigns

Two campaigns were launched: *Thal Bachao Tarla* ('Save Thal Struggle') and 'Don't Desert Deserts'. The second campaign was launched on the eve of the International Year of Deserts in 2006. The first campaign called for the protection of local grazing lands, biodiversity, livelihoods and crafts. Meetings were arranged between local representatives, politicians, intellectuals and concerned citizens. The idea of establishing a Thal Resource Centre emerged from this forum. At the time of writing support was being sought from local government and other stakeholders, but initial activities included the forming of a *Khaddi* (handloom) School to promote hand weaving (Thal's dying art) as a viable livelihood option for poor men and women; planning of a publication to document the Thal desert from an environmental and climate change perspective; and the documentation of the flora and fauna of the Thal (noted above). The second campaign was launched with the wide distribution of two posters and the celebration of *Peelu* day.

Kissan Resource Centres

The project established three *Kissan* (farmer) Resource Centres in the Bela area with the intention to promote climate and environmentally responsive cropping, forestry and livestock practices. The centres were planned to form focal points for the dissemination of technology information to farmers by providing a forum for the various local government extension services to meet with farmers. The centres depend on government support to fulfil their objectives. While delayed by political uncertainty in Pakistan, by the end of the project support had been gained from the local government of Hafizabad, and discussions are under way with the Ministry of Food, Agriculture and Livestock and the Provincial Agriculture Department.

Farmers' festivals

Two farmers' festivals were organized, one each at Thal and Bela. Each festival had a number of activities ranging from exhibitions and farmers' conferences to entertainment such as local dances and camel dances. These festivals were organized at the end of the project as part of the exit strategy. The primary objective was to bring together project beneficiaries, local government officials, civil society organizations and communities who had developed an interest in the project's activities, to share views and experiences. Farmers' conferences were organized to enable local intellectuals and farmers to express their views on threats of climate change, vulnerability to disasters and the role and benefits of the project.

Technology development

The baseline survey allowed existing livelihood patterns and opportunities for diversification to be identified, in turn informing the choice of technologies identified and promoted during the project. As the project focused on rural communities, the technologies predominantly consisted of alternative agricultural practices. Alternative crops were identified in partnership with local experts, government, academic and civil society representatives. Each alternative variety was selected for its ability to withstand drought and flood, and its positive impact on soil stability.

Establishing tree plantations was an early project activity. Around 60,000 tree saplings were planted in Thal and 20,000 in Bela. All the tree varieties in Thal were local species. In the first year, young fruit and non-fruit tree saplings were distributed among interested households. In Thal in particular the survival rate was very low during the first year of the project, with trees only surviving when the household took the time and effort to take care of them. However, in the second year, the project entered into an informal partnership with an existing nursery and older saplings were distributed, resulting in a much improved survival rate. Subsequently, both locals and experts suggested that the quality (and market price) of fruit could be improved if the saplings were grafted, using grafts of improved varieties of fruits. However, grafting is an expensive process and it was not possible to fund this additional activity, leading to frustration among the participating households.

In Bela, lemon and orange fruit trees were planted in existing fields to promote agro-forestry and multiple cropping. The activity commenced by planting the trees on a 4-acre field in Yake Wala and a 3-acre field in Bahar Wala. To meet the demand for young trees generated by the project, the project team sought to set up a nursery. However, a lack of availability of quality seeds and the departure from the project of a team member with experience in setting up nurseries hampered the success of this initiative. By the end of the project the approach of planting fruit trees in cropland had won acceptance in the community. There was a plan to promote a wider range of multiple cropping practices during the project, but ultimately there were insufficient resources to support further demonstrations.

Vegetable farming proved to be one of the key success stories of the project. In Yake Wala, it was noted that landholdings were small and farmers continued to rely on three cash crops: wheat, rice and sugarcane. Women were already active in agriculture but there was found to be great potential to increase their role and enhance their productivity by introducing vegetables on small plots that are available to almost all households. During the initial months, community members, particularly women, were mobilized to grow vegetables. The process was facilitated by distributing good quality seeds among the interested households. At the end of the first season people were delighted to see a good yield of vegetables and the number of interested households increased. By the end of the project vegetable growing had started on a commercial scale in Yake Wala, despite there having been no history of vegetable farming in the village. This activity has contributed to household food security and has helped to diversify livelihood options for farmers who were earlier relying on cash crops, two of which (rice and sugarcane) are water intensive. However, vegetable farming promotion in Thal was not so successful. Reflecting on this failure, the project team identified four differences between the experiences:

- lack of experience in vegetable growing among the Thal field team;
- weather and soil conditions much better suited for vegetable farming in Bela;
- considerable lizard and termite damage in Thal (it was reported by community members that termite numbers had increased following the extinction of bird varieties due to hunting and loss of habitat);

- advice from the local agriculture department was not sought in Thal when it had been in Bela.

In both Bahar Wala and Yake Wala, the women were given training on multiple trades that has helped extend their livelihood options. The training also helped to build the confidence of the participating women. Livestock training was arranged for interested youths from both Thal and Bela, and has subsequently become a readily available and affordable source of veterinary services for the communities. Moreover, three young people have gained an alternative livelihood by selling their veterinary services in Bela.

Finally, an innovative demonstration area was established in Bela. During the project it became clear that in one of the villages the riverside lands are being converted into fields, damaging the natural river ecology and rendering the communities more vulnerable to floods. However, during the relationship-building and awareness-raising process, the communities were unable to understand the importance of conserving the river ecology. The project staff observed that the communities would only be convinced once they had seen the effects for themselves, and developed the idea of a model to demonstrate how conservation of river ecology can offer secure and sustainable livelihood options. A demonstration site was chosen on 50 acres of land adjacent to the Chenab River, filled with features typical of the Bela area. The land was leased by the project, with the aim of demonstrating and raising community awareness of livestock management techniques, organic vegetable farming and the importance of natural resource management (including the provision of medicinal plants, vegetation for grazing, fuel wood and flood protection). However, this experiment could not be sustained owing to a lack of capacity within the project team, lack of resources to maintain a large area of land and, above all, a lack of interest shown by the communities in the initiative.

Resource implications

This three year project had a modest financial resource base of around £89,000 ($131,000) to cover the cost of four field staff and a project manager, office upkeep (three offices), travelling, and the cost of training, technologies, meetings and publications. The staff salaries were lower than the market rate throughout the project period contributing to high staff turnover. This fact contributed to loss of investment in human resources and interrupted the continuity of the project. The project was able to allow team members to participate in training programmes and travel for exposure visits. However, there were insufficient resources to send the staff for paid training opportunities within and outside the country. Moreover, the limited resources to cover staff travel meant that frequent staff meetings were impossible, even though they would have been beneficial in ensuring that experiences were shared. Similarly the project lacked a contingency to cover unexpected losses (such as those due to frosts or termite infestations) and was unable to provide funds to demonstrate grafting.

Materials for the project accounted for £15,500 of the total budget. The project reached approximately 1,000 beneficiaries, resulting in a cost of £89 per beneficiary over three years.

Lessons and challenges

Adaptation to climate change is a relatively new concept for development professionals and organizations, and the experience of this project suggests that staff need to work to build their own capacities before they can be of benefit to the communities. However, as the field teams were drawn from the local population they had a good understanding of the local socio-economic and political atmosphere, and the natural environment. This was a notable feature of the project, as NGOs working in the project areas have been known to employ staff who lack understanding of the local culture and priorities, and fail to build long-term and sustainable relationships with the community. While the use of local employees can lead to research biases, the project team worked hard to ensure that the reporting tools guarded against this possibility.

Gathering local knowledge

Gathering local knowledge was found to be an excellent starting point for an adaptation project since it offers:

- an understanding of communities' perceptions of weather, weather patterns and changes;
- an understanding of the role of the seasons in local culture and how they determine livelihood and daily activities;
- insight into vulnerability to climate hazards;
- an understanding of local or traditional early warning systems and prediction methods.

However, there are limitations to local knowledge which need to be appreciated before conclusions can be drawn. Local knowledge can be highly subjective, subject to exaggeration or misreporting of events. It can be inconsistent: for example, different community members may remember the same event in very different ways. Moreover, there is a risk that researchers may wrongly attribute environmental change to climate change or fail to understand the role of natural variability in weather patterns. Some of these shortcomings can be guarded against through good practice, and continuing the knowledge gathering process throughout the project allows the project team's improving interpretative skill and relationship with the community to be built on. Researchers need to be as well appraised as possible of the nature of climate change, likely climate change impacts, and other local sources of environmental change. It is desirable to complement (and triangulate) findings from local knowledge with information from other sources, such as temperature and rainfall records or ethnographic studies. An understanding of the local culture, language, poetry, slang, songs, celebrations and rituals provides additional depth, as the language frequently contains meaning on many levels and, importantly, oral testimonies are not the only source of local knowledge. Finally, the project found that women are often holders of useful information which men often are unable to provide and, moreover, were found to have a more refined and culturally rich vocabulary.

Social networks

As with the assimilation of local knowledge, community mobilization and organization is a continuous activity that requires strategic thinking, commitment and a readiness to work with the community throughout the project. When planning for social mobilization, it is worthwhile to first identify the existing nodes of social relationships: for example, the traditional institutions of the Thal (story nights, *Lok Sath*, festivals) proved far more effective than modern, introduced community organizations. Similarly, it is important to respect cultural sensitivities. For the communities of Thal and Bela, gender identities are deep rooted and social interaction between women and men is not acceptable. For this reason women and men were organized in all-women and all-men groups in both Thal and Bela.

Throughout this project the awareness-raising activities were recognized as opportunities to build social networks. Exposure visits in particular proved to be an effective tool for orienting and mobilizing the members of the community organizations. The experiences of this project also suggest that it is necessary to allocate funds at the planning stage for community organizations to use in support of local initiatives in line with the project's overall objectives. If the intention is to form independent CBOs then funds should be provided, and deposited in the organization's account, to ensure that seed money for adaptation activities is available.

Awareness raising and training

An unsuccessful training workshop designed to strengthen the organizational capacity of CBOs in Bela offers lessons for similar activities in the future. Owing to a lack of resources, the training was organized in the village. However, the proximity to their homes distracted the participants, who would regularly leave the training sessions to undertake household chores and livelihood activities. Moreover, the approach to communication adopted by the trainers was poorly thought through and failed to respond to the needs of the participants. For example, a session was held on record keeping and accounting, but completely failed to engage the community members, none of whom could read or write. Training sessions organized for local government officials were better received, but did not result in effective networking. A lack of follow up by the project team was blamed for this failure.

Awareness raising in Thal targeted the communities, local government officials and elected representatives at local level and used mechanisms including campaigns, public action (such as peaceful protests and demonstrations) and events that provided opportunities for formal and informal discussions. The success of these activities was reflected in the promise of newly elected local representatives to offer their full cooperation to the project and their invitation for project participants to participate in the drafting of local development plans. However, the long-term commitment of the elected officials is still to be demonstrated. The continued persuasion and pressuring of local representatives at local level will be necessary if the project achievements are to be sustained.

A key lesson from the project was that awareness raising should be understood as a two-way process in which the implementing organization, communities, governments and other stakeholders learn from each other. The local knowledge reports helped cement this process, providing a hitherto missing link between people's perception of the environment and the body of formal scientific work. These reports fulfilled an important role in ensuring that local knowledge is properly documented and publicly available. Finally, video documentaries proved to be an effective tool in communicating with the large section of the population that is not literate. Mass media tools such as television and radio are also effective and should be considered for inclusion in the communication strategy of any adaptation project.

Technologies

A comprehensive picture of existing livelihood patterns, mapped onto current and future climate trends, is necessary before livelihood-centred adaptation interventions can be designed. Such a mapping identifies specific livelihood vulnerabilities and options for building adaptive capacity, and should be drawn up with the different vulnerabilities of women's livelihoods in mind. All livelihood interventions, however, need consistent support over an extended period of time. In Yake Wala vegetable farming started on a commercial scale towards the end of the project. However, many households continued to expect to receive seeds from RDPI: efforts are now under way to link these farmers with an appropriate local government programme. Mechanisms are also required to assess how new livelihood options respond to ongoing climate variability and change. In Thal and Bela, for example, frosts have returned following several years of warm winters, severely affecting crops, vegetables and young saplings. Ideally, projects should build in the ability to respond to emerging climatic conditions. Each of these sustainability issues points to the need to ensure that an adaptation project has a strategy to link communities with government and non-government institutions working on livelihoods. These institutions may include (but are not limited to) microfinance banks, skill development training institutions, and local agriculture, forestry and livestock departments. Moreover, livelihood interventions also need to seek out and develop market linkages, for which resources and specialized expertise are required.

Conclusion

Community organization, awareness raising and culture were central themes throughout this project. Relationship building was integrated into project activities from the start, allowing synergies between the need to learn, educate and network to develop. Baseline research was an opportunity to build trust, awareness-raising activities were used to bring different stakeholders together in non-confrontational settings and the documentation of local knowledge emerged as a chance for the project team and the communities to improve their understanding of climate change. An appreciation of tradition and culture is consistent with this approach and is supported by well-developed relationships between the project team and the communities. In this case, the use of local field teams was a catalyst in this process, as the team members were able to understand and interpret aspects of culture, language and the environment that may otherwise have been missed. As noted, this approach also helped ensure solid and lasting relationships between the communities and the project team and ensured that gender roles were respected rather than an issue for confrontation.

The importance of culture and tradition was particularly clear in the use of the *Lok Sath*, or people's parliament. As a traditional forum this was able to gain acceptance where externally inspired community organizations were not. As a result, the *Lok Sath* was successful in raising awareness of climate change, promoting public accountability, and providing a socially and culturally acceptable discussion platform. The use of story nights similarly capitalized on traditional social forms to help the communities engage with unfamiliar issues. As well as providing an institution for building horizontal community relations, the *Lok Sath* also has the potential – albeit thwarted during the project – to become an important network link between the communities and local government. It is also of interest that the *Lok Sath*, as a socially accepted and locally understood institution, was the forum at which the communities revealed issues to the project team that had not been expressed to them elsewhere. An appreciation of local culture also played a defining role in the selection and promotion of alternative livelihood technologies. The importance of traditional species in providing sustainable livelihoods was clearly demonstrated in Thal, where native tree species and camels were proposed by the project team as methods to cope with the harsh desert conditions. Recognizing a link between traditional livelihood activities (camel rearing and *jaal* tree cultivation), their resilience in the local environment, and traditional festivals (camel dancing and *peelu* gathering) was a significant step, and allowed the project team to uncover both appropriate technologies and an effective approach for their promotion. The festivals provided a mechanism for effecting change from within the communities that tapped into centuries of culture and indigenous knowledge. The broad principle, in which local technologies are well suited to the environment and associated with traditional institutions or events, is one that is likely to be relevant in other contexts.

The support offered by networks that reach outside the community was also important in this project, the more so for the introduction of new practices. Vegetable farming was recognized as an opportunity to make productive use of small plots of land held by community members and was an appropriate activity around which to build the skills and confidence of women. However, the success of the vegetable production initiative was in part due to contact with local government extension officers: in Thal, where this element was lacking, the approach did not take off. Similarly, success in fruit tree cultivation was dependent on partnerships that the project built with local nurseries. The power of demonstrable success to encourage wider take up of new ideas was also evident, both from the increased interest at the end of the first (successful) year of vegetable production in Bela and through the success of exposure visits. Support for farmers was also provided by the *Kissan* Resource Centres in the Bela area, a key network initiative providing access to technology information and a forum for local government extension officers to meet with farmers. However, building relationships requires long-term effort and planning: for example, the need to provide seeds to vegetable farmers beyond the end of the project necessitates an ongoing relationship with government or other service providers, but this step was only taken at a late stage in the project. Similarly, relations built with government institutions can offer support but require effort to sustain, as illustrated by the failure to capitalize on the successful training sessions with government officers, and the need to maintain pressure on newly elected representatives if promises for action are to be translated into improved services for the communities.

Finally, the process adopted was one of mapping livelihoods and climate hazards to identify opportunities to build resilience and diversify livelihood options. This approach principally focused on existing environmental problems, but also identified opportunities for diversifying livelihoods, for example, through multi-cropping. However, the project also offers a stark illustration of the difficulties posed by the uncertainty and variability of climate change, with the unexpected return of frosts severely damaging crops, vegetable and saplings in both Thal and Bela. Adaptation to climate change must also foster resilience against such unexpected events, through diversified assets, livelihood strategies or networks of information and support.

CHAPTER 5

Increasing paddy salinity in coastal Sri Lanka

Prepared with Rohana Weragoda, formerly Project Manager, Practical Action Sri Lanka, currently Senior Programme Officer, World University Service Canada, Badulla, Sri Lanka

Abstract

Temperature increase, sea-level rise and the failure of irrigation systems are all contributing to the increasing salinity of small-scale farmers' rice paddies in coastal Sri Lanka. This project aimed to reintroduce traditional rice varieties to restore yields following the failure of fertilizer-dependent hybrid varieties and neglect by formal research institutions. Critically, farmers themselves led variety-selection research, rebuilding the capacity and confidence to experiment with indigenous varieties that had been sacrificed in favour of the apparent benefits of green revolution technologies. Farmer-to-farmer learning networks facilitated the sharing of skills and experiences, while relationships fostered during the project built networks between the communities, a national seed organization, and the nearby rice research institute.

Introduction

This chapter documents a community-based adaptation intervention carried out in the Hambantota district in the coastal belt of southern Sri Lanka. The villages that form the focus of the work described in this chapter lie along the Walawa River estuary, surrounded by the varied topography of the area which includes marine and estuary foreshore, mangroves, small forests, and lowland and upland cultivation zones. The main road in southern Sri Lanka lies along the coastline, providing the communities with good access to transport and markets, although gravel roads in particular remain damaged following the 2004 Indian Ocean tsunami. Most of the Hambantota communities have access to electricity provided by the main grid. The quality of drinking water is variable; some have access to piped water but many rely on wells that are reduced in number following saline inundation. There is an irrigation system in place for agriculture, but many water channels are poorly maintained. The community is made up of individuals who are mixed in terms of education, occupation and skills. Most of the community members have received primary education, while only around 10 per cent of the community have had secondary education. Only a small percentage of the families have their own savings; instead, most retain assets such as crop harvests, livestock, boats and other equipment.

Cultivation in the district is linked to the monsoon, which normally occurs in two seasons, known locally as *Yala* and *Maha*. *Yala* runs from April to the end of July and receives rain from the south-west monsoons. This is the drier of the two seasons, during which crops such as sorghum, maize, fruit and vegetables are grown in home gardens irrigated by water lifted by private pumps from the river. Dry season cultivation provides an additional income source for the local farmers. From September to February the north-east monsoons feed *Maha*, signalling the onset of paddy cultivation. However, salinization and soil erosion have led to the abandonment of paddy land that lies close to the Walawa River estuary. Widespread saltwater intrusion has reduced yields for many of those farming further from the estuary, while flooding in downstream areas following torrential rain threatens crops, livelihoods and infrastructure. Coastal erosion further undermines the livelihoods of the Hambantota communities, reducing the availability of land and forests that are increasingly – and illegally – relied on by local people as land scarcity increases across the district. Intense rainfall events and rising sea levels are therefore significant determinants of vulnerability in the Hambantota district, and both are anticipated to continue under current climate change predictions.

Salinization of the paddy lands has been principally driven by three factors: saline contamination of irrigation systems, temperature increases and sea-level rise. The Rice Research Institute in Sri Lanka normally reserves its focus for large-scale irrigated paddy cultivation rather than the marginal or small-scale producers such as those in Hambantota. However, without involving farmers the Institute has developed a few varieties which are suitable for saline conditions without the involvement of farmers, but these are limited in terms of the levels of salinity that they can tolerate. Enforcement of local-level environment policies is very weak in the area and as a result the pressure on natural resources is increasing. Changes to the river opening following the tsunami has prevented seawater inflow, providing local people with the opportunity to clear coastal forests and fallow lands for paddy cultivation. However, these clearances include sensitive lands that are important in buffering floods and preventing seawater intrusion. The loss of mangrove and forest cover increases the risks from tsunamis, flooding and coastal erosion and is disrupting the local ecosystem. Given this context, Practical Action Sri Lanka sought to develop a project focusing on the implications of climate change for the existing problems of salinization, flooding and natural resource management. The project worked with 500 households in Godawaya and Walawa *grama niladhari* divisions in Hambantota district over three years.

This chapter focuses in particular on work done to identify traditional rice varieties suitable for cultivation in the degraded paddies through a process of participatory research. The project also worked to extend community and institutional knowledge of climate change, and commenced with an assessment of climate change awareness. This work is reported in the next section, together with a comprehensive assessment of the livelihood vulnerabilities and risks. Next the substantive project activities are discussed, with attention paid to the role of social networks in the development of responses to climate threats. The following section considers the lessons learnt and challenges faced during the project, and the chapter ends with an assessment of the broader implications for community-based adaptation.

Community-based adaptation in coastal communities

Understanding the situation faced by the Hambantota communities was an essential first step towards designing project activities. Ensuring sufficient information was gathered was critical and as a result the first six months of the project were given over to a range of data gathering approaches, as outlined below and summarized in Table 5.1.

Secondary data

Site selection was facilitated through secondary data such as the Sri Lankan census and statistics and International Union for Conservation of Nature

Table 5.1 Data-gathering methods employed to understand livelihoods in Hambantota

Participatory method used	Objective	Information gathered
Key informant discussions	To collect information on natural resources, livelihoods and social and institutional arrangements in the area	Sources of livelihood
Use of natural resources
Existing socio-economic systems
Focus group discussions	To collect background information on resource use, vulnerabilities and capacities to validate secondary and sources	Vulnerabilities
Livelihood system problems
Community resource mapping	To identify the natural resource base and natural resource management methods	Vulnerabilities
Availability and use of natural resources
Area landscape
Risk mapping	To identify risks to livelihoods within the area	Types and causes of risk and threat
Extent of vulnerable areas
Field observation	To further understand biophysical resource vulnerability and existing problems affecting farmers' livelihoods	Risks and threats in relation to biophysical resources
Encroachments
Severity of problems identified
Transect walks	To further understand biophysical resource vulnerability and existing problems affecting farmers' livelihoods	Linking upstream and downstream farmers
Variation in vulnerability

(IUCN) reports on coastal resources. This process revealed the Ambalantota division of Hambantota district in southern Sri Lanka as highly vulnerable to floods, seawater intrusion and natural resource depletion.

Focus group discussions

Focus group discussions were used as a method to initiate dialogue in the villages and collect background information to validate the secondary data. Two focus group discussions involving representatives of the government, non-government and community-based organizations were conducted to discuss problems relevant to livelihoods and to understand resource use and vulnerability. The government representatives were composed of the assistant divisional secretary (who reports to the district secretary, a government appointee, via the divisional secretary), *grama niladhari* (the lowest administrative level) officials from the area, village farmer group leaders, and officers of the Rice Research Institute in the area. This process led to the selection of two paddy-farming communities (from Manajjawa and Walawa) for further vulnerability assessment.

Community resource mapping

Following the focus group discussions, resource and risk mapping were introduced to further understand the vulnerabilities of the two communities. The main objective was to identify the natural resource base and natural resources management approaches. Selected farmer organizations and other community members engaged in drawing maps of community resources, illustrating natural resources such as paddy lands, home gardens, plantation lands, settlements, water bodies, and other important natural resources and sources of livelihoods.

Risk mapping

The groups involved in risk mapping comprised community office bearers (such as the presidents of the village agricultural society and death benevolent society) and the members of farmer organizations. Using the community resource maps, the group members identified the areas at risk from salinization and flooding, water scarce areas and irrigation canals. The main issues identified were flooding, salinity, conflicts between water resource users, and risks due to changing land-use patterns.

Field observations and transect walks

The Practical Action group joined community members and made field visits throughout the villages to complement the results of the focus group discussions and mapping exercises.

Despite the considerable number of tools employed to generate information, the project team felt that other methods could have offered greater detail. Wealth ranking, semi-structured interviews and household case studies would have offered insights into the local economy. If integrated with an investigation of the relationship between income generation and natural resource use, these additional sources could have informed the development of improved coping strategies.

Assessing climate knowledge and awareness

The use of language was critical to understanding local perceptions of climate change. The phrase 'climate change', when translated into Sinhala (*deshagunika viparyasa*), was too broad for the community members to understand: it lacked relevance and meaning in their daily lives. It was essential, therefore, for those facilitating the discussions to have sufficient understanding of vernacular vocabulary to enable climate change to be explored in terms of the experiences of local people. In particular, it was important to have familiarity with local livelihoods, farming systems and the associated language. The project team found that farming communities had developed very precise terminology and employed words with great care. It was only when an appreciation of these nuances had been developed – through the investment of time in the field – that the essence and extent of the local knowledge of climate and weather could be appreciated.

Discussion of local perceptions of climate change took place in groups drawn from within the community. There were no records of any kind held at the local level and therefore the memories of the inhabitants were the only source of information. An appreciation of this led to the formation of groups with participants of similar ages to enable discussions among individuals with similar knowledge. Participants were therefore split into those over 60 years, between 45 and 60 years, and the younger generation (below 45 years). It was also noted that women within the communities were frequently closely connected with the environment through activities such as firewood collection, farming and homestead vegetable gardening, and as a result were often more sensitive to changes in resource availability and temperature than men. Community groups were therefore selected to ensure representation of farming activity, age groups and gender.

Knowledge collection was facilitated through the employment of participatory processes including group and focus group discussions, resource and risk mapping, and field observations. However, one inevitable consequence of mixed groups was the diversity of views and therefore the time required to develop a consensus position. In this context even the six months allowed for data capture proved to be a challenging timescale. Alongside community groups, local professionals such as schoolteachers, local administrators and environmental committee members were involved in scoping climate change awareness. These groups are particularly significant as they were the most likely to have had exposure to scientific thinking.

The information gathered by the project revealed that the community had access to modern weather information. However, traditional methods remained significant. At the time of the project study, farmers continued to rely on a forecasting system known as *Litha*. The *Litha* is a table prepared by an astrologist, derived from the phases of the moon and the positions of the stars and planets. The table is referred to when farmers are planning their activities, determining, for example, when rain will fall, when seeds will germinate, or the best time to plant crops so as to avoid pest attacks. A full moon, for instance, is associated with rain. However, while respondents reported the *Litha* to still be in use, it was judged to have become less effective in recent years. While this finding may suggest that climate change is undermining traditional techniques, care is required in interpreting this data. Further research would be necessary to establish whether, for example, this perception was due to the increasing reliability and/or prevalence of modern forecasts.

Alongside the *Litha*, there are a variety of traditional forecasting indicators that the farmers use to interpret their observations. Examples of these are given in Table 5.2, but it should be noted that, as with the *Litha*, the effectiveness of these methods was being debated within the community. It was also reported that the traditional techniques are not being passed on to the younger generation, who are increasingly reliant on modern farming and forecasting practices.

Most of the modern farmers reported using both traditional and modern sources of information for decision making, although the older generation continued to prefer local knowledge systems. Discussions revealed that the older age group perceived no particular advantage in using scientific sources. While it was acknowledged that traditional methods offered far from perfect predictions, the elder community members were also aware that modern forecasts vary in accuracy.

Table 5.2 Examples of traditional forecasting indicators

Observation	Prediction
Large termites start breeding during a dry period	Rain will come soon
Ants appear with their eggs and move to a new nest	Rain will start within 24 hours
Small termites start breeding during a rainy period	Rain will stop soon
A noise is heard emanating from the sea	Rain will come within seven hours and last for seven days
Off-season trees such as tamarind and wood apple give good yields	Good future rainy season – farmers cultivate large areas
Dogs and cattle make unusual sounds	Destructive rainy season leading to disasters is anticipated

Social networks

Most of the families in the area were found to be members of either the fishery or the farming society at the start of the project. Farmer organizations in each village act as the decision-making forum for issues such as the farming calendar, water distribution and seed selection. These organizations are also linked with government agrarian services centres, while the fishery society is linked to the Ministry of Fisheries through the fisheries inspector. In addition to these main groupings, there were also small societies and committees that had been established following the tsunami by different organizations to enable relief and reconstruction work. Depending on their occupation, farmers also reported being involved in informal institutions such as the *Yaya* (field) groups and *Seettu* (community-managed revolving fund) groups.

A consensus was reached during the participatory assessment process which established floods (crop losses) and salinization (low yields and crop damage) as the main threats to livelihoods. The community expressed a clear desire to reduce the risks of crop failure and unprofitable yields, and consequently the project developed a strategy to build capacity around resource use and management. In particular, this approach sought to establish and replicate best practices in crop cultivation. The main vehicle for training was the paddy farmer group, which was mobilized to adopt and develop sustainable practices in their farming. The paddy farmer groups required no additional incentive to become involved in the project: they recognized that it offered an opportunity for them to improve their harvests and income. However, although farming activities are shared between men and women, the group was dominated by male participants. Investigation by the project team established that rather than there being a formal bar on women participating, their time commitment to household activities was an obstacle to attendance. The project made efforts to accommodate women by finding alternative meeting times suitable for rural families, such as evenings and weekends.

The paddy farmer group consisted of 16 farm families, cultivating about 35 acres (14 ha) of paddy lands. The paddy fields yielded an uneven crop, with variation within and between fields. The extent of the salinity problem was such that at the time of the project some fields were carrying no crops at all. The farmers had been using saline-resistant varieties that had been developed at the Ambalantota rice research station (such as AT 362 and AT 354), which had been found to grow with some success in the fields. However, in the badly affected areas the resistant varieties were still failing. The paddy farmer group in Manajjawa (the upstream farmers) were particularly affected. The only water supply for this area was from an upper field channel that was regularly contaminated with salt water. During the year, this water circulates and repeatedly irrigates the paddy, with the result that the salt concentration rises.

The farmer group agreed to trial traditional rice varieties alongside the research station varieties. The trial was supported by the National Federation for Conservation of Traditional Seeds and Agricultural Resources (NFCTSAR), a non-governmental organization, which supplied the farmers with traditional seeds that had been saved and multiplied (local seed banks had fallen out of use following the introduction of high-yielding hybrids in the 1960s and 1970s). After seeing that some varieties were performing well in the saline soils the farmers developed confidence in their ability to research and select varieties. Moreover, with the help of the project team the community were able to form links between the farmer organizations and the Government Rice Research Institute, yielding advice and soil testing. Prior to developing this network the role of the institute had been simply to provide seed to the farmers with little or no interaction. This situation was reversed during the project, as lines of communication were established between the research institute and farmer organizations. The project also facilitated contact with local government via the *grama niladhari* (at the village level) and the irrigation department, prompting important actions such as sluice gate renovation (to prevent seawater flowing into the paddy fields) and support for improving irrigation to ensure a supply of fresh rather than saline water. Critically, the farmer groups also extended their network to other farmers in the area. This process of farmer to farmer learning was a major development during the project. Those testing traditional varieties shared their findings with farmers from the adjacent village who were facing similar problems. This process stimulated a discussion between farmers on variety selection, increased the knowledge of neighbouring farmers on climate change issues, and enabled the farmer groups to become change agents at the community level. By the end of the project a traditional paddy cultivator group had been established and was formally attached to the NFCTSAR, providing a wider network of contact and support for the Hambantota farmers and helping sustain the project developments once Practical Action had withdrawn.

Technology development

The principal technological aim of the project was to initiate a farmer-led process of rice variety selection. Participatory variety selection was conducted to determine the acceptability of and preferences for the traditional varieties under saline conditions in Hambantota District. The approach helped farmers to develop their capacities in the selection of varieties that are appropriate for the local conditions. Formal rice research in Sri Lanka is focused on the main irrigated paddy cultivation sector and is limited in its relevance to marginal or small-scale paddy farmers. Improved varieties are frequently introduced by the research stations, but are not always suitable for use over the long term because of the continuously changing planting environment. The formal research process thus fails to focus on the needs of the farmers, does not build the decision-making capacity of farmers, and ignores local knowledge in the selection process. Participatory variety selection aims to address these shortcomings, and was the farmers' first opportunity to select traditional varieties.

Variety selection was conducted by the farmers, in their own fields, during the 2005 *Yala* season at Manajjawa of Ambalantota Divisional secretariat of Hambantota District. Initially 16 progressive farmers were involved in variety selection from 10 traditional varieties. Selections were made on the basis of scoring each variety on a range of qualitative indicators. Each variety was given a mark between 1 and 10, based on the farmer's preference, where 1 was the best available score. The criteria were established by the farmers and assessed plant height, duration of the crop, grain quality, grain colour, saline tolerance and the grain yield. The farmers planted up to 5 kg per variety in the saline-affected areas of their paddy fields and continuously observed the growth and changes in the plants up to harvesting. Two farmers out of 16 cultivated all 10 varieties while the rest cultivated three or four varieties based on their preferences after seeing how the 10 varieties performed in the first trials.

The variety selection process was supported by activities to help improve soil quality (much degraded in the saline-affected areas) and soil water retention. Training and support was provided to enable farmers to produce organic manure (from treated rice husks and compost) and bio-pesticides (such as neem). As noted above, knowledge sharing took place during the trial. Farmers who cultivated saline-affected lands in the surrounding area were invited to the field during crop growth to make their own observations and discuss progress with the farmers involved in the trial. The participatory variety selection process enabled needs-based selection of paddy varieties by the farmers and helped promote quicker adoption of useful varieties in the farming community. Moreover, the traditional rice varieties carried a premium price at market, returning 60 to 80 Sri Lankan rupees (US$0.54–0.72, exchange rate 1 rupee = $0.009, 10 December 2008) per kilogram compared with 40 to 45 rupees per kilogram for the hybrids.

The project also sought to respond to concerns over costal erosion and the associated loss of land resources. Coastal vegetation is the principal means of arresting erosion, with the roots of appropriate coastal species anchoring the sand. The community groups were therefore keen to receive seedlings, but lacked the necessary knowledge of coastal land management and species selection. The project organizers facilitated a process of learning for the community, in which experts were invited to talk to the community in workshops and during field visits, improving local knowledge of coastal vegetation and its importance in combating climate change-induced sea-level rise. To facilitate the ongoing supply of suitable vegetation, community members established nurseries under the guidance of the project, with the intention of supplying plants to a number of different planting projects. Following training, three nurseries were established, with payment provided at an agreed rate per plant. Community organizations were formed to manage planting and upkeep, and 18,700 plants were successfully introduced into the coastal area. However, 75 per cent of the plants subsequently perished during the dry season owing to a failure on the part of a local partner organization to distribute upkeep funds to the community organizations. Practical Action cut ties with the partner organization following this incident.

Awareness raising

The discussions and interviews conducted at the start of the project identified that the awareness on climate change was minimal, both among community members and external stakeholders such as irrigation and agriculture professionals, environmental managers, NGOs and the government. Awareness creation programmes were therefore conducted at different levels, involving community groups, schoolchildren and the decision-making members of key organizations including those at research stations, agriculture departments and divisional secretariats.

Awareness programmes for the communities and schoolchildren used multimedia presentations with pictures, depicting the resource depletion, soil erosion and photos of extreme events. Explanations with graphics were helpful in generating understanding of the anticipated impacts of climate change. A series of discussions were carried out with the local-level governmental and non-governmental organizations to establish their views on climate change, leading to workshops conducted with the support of Sri Lanka Centre for Climate Change and involving community members, agriculture and environment professionals, government representatives and NGOs. The focus was on agriculture, with particular attention given to how climate variations change cropping patterns and the yields, pests and diseases of different crops. The discussion was also focused on the impacts of emerging rainfall patterns and disaster management following extreme events.

Resource implications

The project team funded the provision of seed for the initial research and was responsible for providing technical advice (such as the production of organic pesticides and organic manure preparation) and assistance with building relationships with the Rice Research Institute and the Department of Irrigation. Staff were drawn from both Practical Action Sri Lanka and local partner organizations, as detailed in Table 5.3.

Table 5.3 Project staff levels

Staff member	Commitment
Project manager, Practical Action	Full time
Project officer, Practical Action	Full time
Field officer (coastal vegetation development), local partner organization	Full time
Field officer (variety selection), local partner organization	Full time
Secretary, Practical Action	4 months
Secretary, local partner organization	4 months

The project served 800 direct beneficiaries, with a further 3,500 individuals who indirectly benefited by attending workshops or other similar project activities. The total project budget was £67,000 ($99,000), yielding a cost per direct beneficiary of £84 over the three years of the project.

Lessons and challenges

In addressing vulnerabilities, stakeholder participation is essential so that the local situation can be assessed and coping mechanisms identified. Communities have their own observations of climate variation and environmental change at the local level that must be appreciated when developing programmes for adaptation. However, in the assessment process conducted during this project a wide range of stakeholders were identified and participated throughout: the agrarian service department, the irrigation department, and the Rice Research Institute in Ambalantota Divisional secretariat area; non-government organizations; community-based groups; and community leaders. While farmers had sufficient coping mechanisms to survive in the short term, it was apparent that collective action among all stakeholders was critical if the community was to adopt the new strategies necessary to survive in an increasingly challenging environment. The status quo had been failing the farmers: the local governmental bodies exist to help, yet even the Rice Research Institute, situated within 5 km of the paddies, failed to provide a service to the farmers or take the initiative in addressing the salinity problems.

The wide-ranging assessment exercise identified the critical problems in the area to be salinity, irrigation and a lack of alternatives for the farmers. The use of participatory tools such as risk mapping was effective in encouraging collaboration among the stakeholders. The coordination among the institutes and stakeholders facilitated by the project was essential in identifying solutions: soil testing and options for addressing salinity were provided by the research institute; the irrigation department renovated sluice gates and irrigation systems; and the identification of traditional saline tolerant varieties was coordinated by the National Federation for Conservation of Traditional Seeds and Agricultural Resources.

A key contribution of the project was in raising awareness of local solutions to climate change threats. Forgotten varieties of indigenous rice were demonstrated to be able to offer a solution to the increasing soil salinity. Overall there are around 2,000 traditional rice varieties in Sri Lanka, many of which are very high in nutritional value and have medicinal properties, are resistant to extreme drought conditions, particular diseases and pests. Cultivating these traditional varieties has helped marginalized, salinity-affected farmers to cope, while the collaborative approach adopted by the project has had a positive impact on the attitude of local agricultural institutions. The participatory approach to variety selection proved particularly beneficial, overcoming the limitations of the conventional research system in meeting the needs of marginalized farmers and integrating local knowledge into the selection process.

In certain cases, the development of coastal vegetation by community groups was not successful as there were no visible short-term benefits. The communities engaged in activities such as developing nurseries and planting coastal trees, but only until the end of the project. As with other natural resource management activities, long-term commitment is essential, yet follow-up activities were needed to ensure the ongoing support of the community. Awareness raising, training and planting programmes helped the community to develop environmental conservation understanding and practices. However, a lack of knowledge and coordination on coastal vegetation among the communities and local organizations limited the successful establishment of green belts. Moreover, 25 per cent plant survival is not abnormal in the difficult coastal conditions, making it important to manage expectations and adopt low-effort planting techniques.

A particular difficulty faced by the project resulted from the prevalence of non-governmental organizations in the area. Community members were reluctant to participate in adaptation meetings as they had already had many societies established to support post-tsunami activities. For example, in one village there were 33 societies formed by different NGOs. As some of the organizations provided money for participation it was very difficult to convince the communities to participate in climate change awareness workshops or activities. For example, while developing coastal vegetation belts for environmental protection reduces erosion and seawater intrusion, and thereby directly benefits communities, many people were attracted to the short-term material benefits that they could receive from other organizations. This can also reduce social cohesion, as competition developed among families over the benefits that they were able to accumulate. School awareness programmes, however, were significantly more successful, and were conducted with the support and participation of schoolteachers and children.

Finally, a lack of national policies for promoting adaptation, the low priority for climate change work among development organizations, and the low level of clarity around climate change impacts posed fundamental and ongoing constraints to the development of adaptation programmes in Sri Lanka.

Conclusion

The participatory research approach adopted during this project demonstrates the importance of experimentation for adaptation. Participatory research provided the farmers with a supportive environment within which they had access to the resources necessary for experimentation and were able to demonstrate the efficacy and efficiency of research that is locally informed and farmer led. As a marginalized group, the Hambantota farmers were simultaneously experiencing the impacts of climate change while being sidelined by formal institutions. Prior to the project, research was dominated by the needs of others (large-scale rice growers) and had proved to be of limited usefulness to the local farmers. By building a network of relationships between small-scale growers, formal research institutions, government extension services and a national NGO, the project created an environment within which the farmers were able to assess the threats to their livelihoods and define their own response. Underpinned by collective action through this network, the principal technological adaptation was a soft technology: the capacity of the farmers to use experimentation to find solutions to their own problems and the confidence to apply the same technique to future challenges. The results of farm-based research were both appropriate to the specific local conditions and immediately applicable by the farmers themselves. Horizontal, farmer-to-farmer relationships enabled immediate dissemination of the research through a process of shared experiences, in which individuals from outside the project were able to interact with the experiments and experimenters.

The work undertaken in this project also underlines the importance of biodiversity to adaptation. It was only through the preservation and free availability of a diversity of local varieties that the farmers of Hambantota were able to assess and select seeds capable of surviving in the degraded environment. While preservation of biodiversity is central to adaptation, the manner and location is also significant. An important contribution of the project was linking the local farming community with the NGO National Federation for Conservation of Traditional Seeds and Agricultural Resources. Awareness of, access to and the knowledge to harness biodiversity were all necessary ingredients for success. Significantly, however, awareness, access and knowledge had all been lost with the demise of local seed banks following the introduction of high-yielding hybrids. As climate change pressures mount, this experience is a warning against sacrificing diversity for other supposed gains. Local knowledge of seed attributes is now being rekindled through participatory research, but the failure of externally defined varieties to meet local needs illustrates the importance of (preferably on farm) variety diversity and common ownership of genetic resources into the future.

Social networks played a significant role throughout the project. The relationships built to achieve participatory research were a particular success and capitalized on the pre-existing farmer groups within the community. Without building contacts between these groups, the government institutions and the national NGO, the project would have made no progress. However, translating biodiversity into resilience to climate uncertainty will require sustainable access to these institutions, the more so as long as seed banks remain off farm. The relationships developed through projects such as this must also be transformed into a sustainable network if adaptation is to be achieved in the long term. It is also important to examine the limits of what is achievable through these networks: while capacity on climate change issues was built within local government institutions, the larger problem of the lack of a national adaptation plan was not directly addressed. Similarly, while lack of relevant national research on climate change impacts was identified as a fundamental barrier to adaptation, change in this area was not advocated. Yet the connectivity developed through the project, both with other farmer groups in the area and with more distant stakeholders, could have been exploited to explicitly address policy-level issues alongside immediate needs. Analysis of the policy framework in addition to comprehensive data gathering at the initial stage of the project would have supported such an approach, albeit with an increase in time and resource overheads.

CHAPTER 6

Increasing drought in arid and semi-arid Kenya

Prepared with Cynthia Awuor, Research Fellow, African Centre for Technology Studies, Kenya

Abstract

This chapter describes a project in one of Eastern Kenya's arid and semi-arid areas in which communities are faced with increasingly unpredictable rainfall and frequent drought. The project set out with three aims: to increase food security through improved livelihood resilience and reduced vulnerability to drought; to reduce poverty through improved livelihoods; and to integrate climate change into policy development. By employing an approach that engaged government, community and meteorological science stakeholders the project was able to integrate forecasting into agricultural practice, develop approaches for addressing unreliable water supply, and ensure the sustainability and widespread dissemination of the project outcomes.

Introduction

Many parts of Kenya are already experiencing unpredictable weather, including strong winds in coastal areas and more frequent droughts that are often followed by floods. In a country in which around 80 per cent of the land is classified as arid and semi-arid, climate change poses particular threats to water availability, agriculture, food security, human and animal health, and biodiversity. Moreover, significant environmental changes could disrupt transport and telecommunications infrastructure while also threatening tourism, a major source of employment and of foreign exchange earnings. Adaptation to climate change is thus crucial for the achievement of sustainable development. Although many Kenyan communities have been long used to adapting to variations in weather and environmental changes, the additional threats and stresses caused by climate change mean that effective, context-specific and timely adaptation has become and will continue to be more vital.

This chapter focuses on the experiences of a community that is adapting to increased drought in one of Kenya's arid and semi-arid districts. The community is located in the Sakai sub-location, Kisau Division, Makueni District, in Kenya's Eastern Province. Sakai covers an area of approximately 25 square kilometres and has a population of around 4,800 people who mainly conduct small-scale, rain-fed agriculture and livestock rearing. The area experiences two rainy seasons: the long rains (March to May) and the short rains (October to December). According to area residents, both the long and short rains seasons were reliable up to the 1970s, and the community used to plant and harvest twice a year. However, since the 1980s, the long rains have become unreliable, leaving the community with one dependable annual harvest.

A situation review before the project began revealed that drought is the most prevalent disaster affecting Kenya. The arid and semi-arid areas are particularly susceptible to drought and pose a huge challenge to the developmental plans of the country. These areas are home to a substantial portion of the total population (around 25 per cent), and while livestock keeping is a key economic activity in the arid and semi-arid areas, it contributes only a negligible amount to economic growth of the country. Vulnerability to climate variability and change in the Sakai is exacerbated by the community's heavy reliance on crops that are very sensitive to drought. In addition, natural resource degradation, inadequate infrastructure and provision of social services are key challenges facing the area. So the need for basic development underlies the additional need for adaptation, and adaptation is necessary to reduce vulnerability and enhance resilience in response to observed or expected changes in climate and their impacts.

Major manifestations and impacts of climate variability and change in Sakai include early onset and cessation of the rainy seasons, frequent and prolonged droughts, frequent and increasingly severe water shortages, and famine. These lead to:

- household food insecurity which manifests itself as hunger, food rationing, poor nutrition and ultimately starvation;
- household conflicts over access to reduced quantities of food;
- inadequate water for both household and farming activities;
- inadequate fodder and pasture for livestock;
- high food prices and low livestock prices (the latter drives down incomes from livestock businesses);
- children leaving school to help search for water or to sell livestock;
- people and animals becoming more vulnerable to diseases because of poor nutrition;
- development activities stalling because the little income that is available is spent on food.

These challenges are expected to worsen as climate change progresses. In view of this, a pilot project aimed at increasing the community's resilience to drought was initiated in Sakai in 2006. The Sakai pilot project is part of a regional project on 'Integrating vulnerability and adaptation to climate change into sustainable development policy planning and implementation in southern and eastern Africa', funded by the Global Environment Facility (GEF) through the United Nations Environment Programme (UNEP), and the governments of the Netherlands and Norway. The African Centre for Technology Studies (ACTS), International Institute for Sustainable Development (IISD) and UNEP provide technical oversight to the regional project. The Sakai pilot project is being implemented by the local community together with the Centre for Science and Technology Innovations (CSTI), and the Arid Lands Resource Management Project (ALRMP, a government body that is donor funded).

Community-based adaptation in arid and semi-arid areas

The CSTI and ALRMP teams commenced the project by conducting a review of existing literature on physical, environmental and socio-economic issues in Makueni district. The Makueni District was selected based on its size, high population density, unique dryland characteristics and the devastating impacts of past droughts on the local communities. A reconnaissance study was conducted to identify the most suitable area for project implementation based on various criteria, including vulnerability to drought, existing local institutions and organizational structures, and the community's willingness to actively participate in a project. During this study, the project team reviewed documents produced by ALRMP on their ongoing projects in the district as well as reports by relevant government ministries. These provided background information on biophysical, demographic and socio-economic factors present in the district. Based on the outcomes of the review, a consultation meeting was held with various stakeholders from the district including government representatives, local administration and community leaders. From the discussions, it was agreed that Sakai sub-location would be suitable for project implementation.

Sakai sub-location was judged a particularly suitable site for an adaptation project for a variety of reasons: it is representative of the other locations of Makueni District in rainfall pattern, soil type, crops and vegetation cover; it has a high concentration of population and thus provides an opportunity for maximizing project impacts; the community's livelihood activities have high sensitivity to drought; there is relatively easy access to the area; there is potential for upscaling project interventions; and there are ongoing related activities in the area, such as drought management activities of ALRMP, among others. In particular, the community (through their leaders) had expressed interest in engaging in project activities that would enable them to improve their livelihoods and their ability to cope with droughts.

Further research was conducted to collect baseline physical, demographic and socio-economic data to establish the characteristics, needs and priorities of area households in the face of drought. A questionnaire was developed and face-to-face interviews conducted among 75 randomly sampled households in the sub-location. Data on demographic characteristics were gathered, including age, gender, household sources and levels of income, patterns of expenditure, food situation (in terms of variety, availability and adequacy), land use characteristics (including common crops and their proportions on agricultural land), access to and availability of water (including major sources of water for domestic and agricultural use), main sources of fuel, existing climate forecasts and weather information, household production systems, main sources and quality of seeds, incidences and impacts of past droughts as well as coping strategies, and other environmental problems such as soil erosion. The mean and mode were calculated for a variety of variables to form a basis for the analysis.

The household surveys revealed that farming was the major source of income, followed by casual labour. Those who relied on farming as a source of income depended on it entirely. Five per cent of households depended on remittances for a large proportion (80 to 90 per cent) of their incomes and were thus vulnerable to large-scale shocks such as climate change-induced droughts because the remittances are not consistent and depend on the capability and goodwill of the remitters. About 10 per cent of the households engaged in small-scale businesses that provided about 50 to 75 per cent of their incomes. In terms of household expenditure, 70 per cent of the residents spent between 50 and 98 per cent of their incomes on food (and more than 54 per cent did not have enough to eat). Seventy-two per cent of the households spent between 5 and 10 per cent of their incomes on health; 51 per cent spent between 10 and 30 per cent of their incomes on education. Expenditure on personal items was also limited to between 5 and 10 per cent.

In terms of availability and access to water, the most common sources for domestic use during the dry seasons are rivers or streams, wells, boreholes and sand dams. For nearly 90 per cent of the households the major source of water during the dry season is streams or rivers (see Table 6.1). Most of the rivers and streams in the area are seasonal; their water levels depend on rainfall amounts. The community's vulnerability to water shortages has therefore been exacerbated as the frequency and intensity of droughts has increased. Although during the wet season rainwater harvesting is a significant source for more than 46 per cent of the households, rivers or streams remain the dominant water sources. Wells account for much less water even during the wet season. Generally, the average round trip distance to watering points is approximately 4 km during dry spells.

Table 6.1 Major sources of water for domestic use

Source of water	Number of respondents	Percentage of respondents
Rivers/streams	67	89
Wells	6	8
Boreholes	2	3
Total	75	100

The household questionnaire provided first-hand information about the community, and the process facilitated a strengthening of the relationship between community members and the project team. Analysis of the data provided the project team with good information on priority needs as well as gaps that the project could fill. Key problems identified by the community though this process included: unpredictable weather (especially rains); frequent droughts that exacerbate water shortages, poor water quality, crop failure, loss of livestock and famine; high cost of good quality agricultural seeds; insufficient wood fuel for domestic use; poor quality roads that negatively affect access to markets and poverty. To address these problems, the community and project team held a series of consultation meetings and group discussions involving stakeholders such as government representatives from various ministries, local community members, the project team and representatives of the Kenya meteorological services. The community was involved in prioritizing their problems and identifying key gaps in current drought management activities. Throughout this process the community and project team identified feasible interventions and desired outcomes.

The pilot project was therefore developed with the following goals:

1. Increase household food security through increased livelihood resilience and reduced vulnerability to drought.
2. Reduce poverty through improved livelihoods.
3. Facilitate integration of climate change and adaptation into policy development and planning.

A number of activities were planned in consultation with the local community. Through them, the project sought to demonstrate how to enhance resilience to drought and reduce vulnerability to climate change, reduce poverty and improve livelihoods.

Assessing climate knowledge and awareness

During the collection of baseline information in Sakai, data on local knowledge of climate and weather was collected. About three-quarters of the respondents were aware of indigenous or traditional methods of forecasting rainfall, including traditional weather indicators; 41 per cent were receiving weather information from traditional sources, including traditional weather forecasters. Of these, 28 per cent used this information for seed selection; 36 per cent for tilling, terracing and repairing agricultural land; and 29 per cent used it for planting. Almost all (88 per cent) of the respondents reported receiving weather information from other sources including radio, television, newspapers and agricultural extension officers, and around two-thirds acknowledged that they received weather information in good time. It was noted that additional information such as dates of the onset, cessation and duration of rains as well as suitable crop and seed varieties for any given season would be useful.

While traditional sources of weather information have been useful, and are widely accepted among the community, it was noted that they are not adequate in terms of providing medium- and long-term climate predictions. In addition, they have not provided sufficient detail to enable community members to plan and sustain their agricultural activities. The project endeavoured to fill this gap by augmenting traditional weather information with scientific weather forecasts. The team downscaled scientific weather forecasts for the sub-location, and communicated this information in agricultural terms. Regional weather forecasts obtained from the Kenya Meteorological Department were regularly being downscaled for the sub-location by the meteorologist on the project team. Once the seasonal forecasts had been prepared, the team's meteorologist worked together with the agricultural extension services officer (who is a member of the ALRMP team) to identify suitable crop and seed varieties. It is hoped that this initiative will outlive the project, as ALRMP is a government initiative that includes the Kenya Meteorological Department. Having seen the benefits of the approach, it is anticipated that the Ministry of Agriculture will continue to work with the Meteorological Department.

Community members had been trained on appropriate use of weather information and were regularly provided with information detailing the expected dates of the onset and cessation of rain, duration and amount of expected rainfall, suitable crop and seed varieties for a given season, and dates for land preparation and sowing, among others. The information was communicated using methods such as village meetings, radio, local newspapers, brochures, newsletters and cropping calendars. This approach has been successful and the forecasts were consistently fairly accurate. During one season, crops were attacked by army worms as a result of excess moisture in the soil following high rainfall early in the season. The community incurred some losses but learnt the need to be better prepared for unexpected changes. The concept of probability-based weather forecasts was not well understood by the community. However, in identifying suitable crop and seed varieties across seasons, provisions were made to vary the diversity of seed varieties so that if the main varieties planted in the season failed the community did not incur a complete loss (since other varieties survived).

While timely, appropriate and comprehensive communication of weather forecasts has been useful, there is need for data on medium- and long-term climate predictions for the area. This would be useful in preparing the community for likely changes in future climate. It is envisioned that further strategic collaboration with the Inter-Governmental Authority on Development (IGAD) Climate Predictions and Applications Centre (ICPAC) and the Kenya Meteorological Department would be instrumental in gathering this data. While climate change predictions are currently uncertain, with continuous research and improved quality, it is envisioned that medium- to long-term impacts will be identified with increased certainty. The intention is to provide the community with this information in an appropriate manner to help them prepare better for medium- to long-term impacts.

Awareness raising

The baseline data revealed that almost all the households recognized two seasons of rainfall as their main planting seasons. They identified increased frequency and intensity of drought, early cessation, irregular distribution, and delayed onset of rains as common problems in the area. This awareness provided a good entry point for the project team to create awareness about climate change, and its potential impacts on the area.

The project team informed community members about the changes in weather that are taking place globally. Examples of climate change-related impacts were drawn from global news about climate-related disasters (especially droughts, floods and windstorms) that had occurred in other parts of Kenya as well as in other countries and regions and had been aired through radio and television. The community was informed that, in Sakai, the observed changes with respect to frequency and severity of droughts and their impacts were likely to continue and possibly worsen with time. This was done through the focus group discussions and during the initial workshop on the use of weather information for agricultural planning. The methodology was participatory in that community members were asked to provide examples of disasters that they had heard about as an entry point to creating awareness about climate change and its impacts on their livelihoods and development. Although the community was aware that climate is changing, the attribution of these changes to greenhouse gas emissions remains weak. Climate change science has not yet been well understood at the local level and there is need to explore ways of better communicating climate science to community members. Use of audiovisual and vivid means of communication such as drama, music, art and poetry could be more effective.

Networking and policy influencing

A review of national policies on disaster management and sustainable development of arid and semi-arid lands showed that they do strive to improve the welfare of the vulnerable groups. However, insufficient political will, human resources and organizational capacity for implementation, and socio-cultural factors such as the reluctance of communities to embrace new livelihoods, have made it very difficult to translate these policies into practice.

It is important to integrate climate change considerations into relevant policies. The promotion of adaptation measures can be achieved through awareness creation, research and relevant and targeted capacity building. One of the pilot project team members from ALRMP received training on translation of research into policy by the International Development Research Centre Climate Change Adaptation in Africa Programme in September 2007. The intention is for this new skill to be used to draw lessons from the field and appropriately package and communicate them to policy makers in a bid to promote implementation of adaptation measures.

In terms of national policy influence, the Arid Lands Resource Management Project has facilitated the integration of climate change adaptation into the National Disaster Management Policy, currently under review. A draft of the document has been tabled in parliament, and the outcome of parliamentary deliberations on it is awaited. Moreover, information on the pilot project is regularly updated on the project website (www.csti.or.ke). Several papers have been prepared based on the project's experiences and presented at various national and international meetings. A project brochure and a video documentary showcasing field-level activities have been produced and distributed by the project implementation and management teams to various audiences including national policy makers (particularly from the Ministries of Environment, Agriculture, Social Services and Water), international policy makers at the 13th United Nations Framework Convention on Climate Change Conference of Parties (UNFCCC-COP) and other climate change adaptation practitioners in Africa. The video documentary has been distributed by the same teams to parliamentarians, the local community (to facilitate peer learning) and other international organizations. Policy briefs highlighting the importance of integrating climate change adaptation into sustainable development policy planning and implementation, drawing on lessons from the pilot project, are to be produced at the end of the project.

At district and divisional levels, relevant policy makers from the Ministries of Environment, Agriculture, Social Services, Water, Planning and National Development and Finance have been engaged in the project through the Arid Lands Resource Management Project. These stakeholders were linked by the project to the community in a relationship that, as it is rooted in ALRMP, is anticipated to continue beyond the life of the project. The project's interventions are currently being upscaled to 28 other districts where the ALRMP works. This is a key success that has the potential to produce positive impacts on a larger area of the country.

Technology development

In addition to downscaling, appropriate interpretation, packaging and timely communication of weather forecasts to the community members, project interventions also included training of community members on appropriate agricultural practices and animal husbandry. Such practices included appropriate seed sowing methods (including the correct depth and spacing of seeds), methods and timing for weeding and dressing, crop rotation and mulching. Training in animal husbandry included paddocking, de-worming, cattle dipping and zero grazing. These practices have helped to improve agricultural outputs through better farm management. Farmers were also trained in the identification, retrieval, selection, bulking and storage of good quality local and hybrid seeds. Farmers were trained on how to select the best seeds by observing various characteristics such as the size, colour of the seed, texture of the seed coat, presence or absence of defects and whether the seed floats or sinks in water. Experts (agricultural extension officers) trained the community on selection of seeds during the first harvest. During subsequent harvests, the community members selected quality seeds individually and the experts examined those selected to verify their quality. Storage was facilitated by dusting seeds in ash and storing in waterproof bags in cool and dry places (on wooden racks that are not in contact with the ground). In addition, field-based training was provided on pest control, post-harvest storage and management. Seed bulking and saving was not practised by the community before the project began, but training and continuous advice by the extension officer has embedded this practice. Initially, hybrid seeds were sourced from the Kenya Agricultural Research Institute. Seeds for traditional drought-resistant crops such as gadam sorghum, pearl and finger millet, green grams and cowpeas were sourced locally. Green grams and cowpeas had already been in use in the area, but the other varieties had long been abandoned in favour of maize and beans. The project re-introduced them and sought to popularize these traditional crops. Drought-resistant maize hybrids were also introduced, but using open-pollinated varieties, the seeds of which the farmers were able to save for the following year. Through seed bulking, farmers are now able to produce quality seeds locally. From time to time, the project will assist by purchasing quality hybrid seeds upon request by farmers. During the project, inputs were provided to the community including chemical pesticides, molasses and polythene bags. Locally available, low cost inputs such as ash were promoted as part of the training.

Demonstration sites were established among 40 households, who were also provided with good quality seeds, to show the benefits of use of weather information in agricultural planning as well as application of proper agricultural practices. These farmers have been conducting farmer-to-farmer training, and distributing good quality seeds from their farms to other farmers in the area. Many of the farmer-to-farmer relationships were in existence prior to the project, but have been strengthened through the project by ensuring that the 40 selected farmers committed to support at least two neighbours each. Project interventions have been upscaled quite well in the area; about 80 per cent of the farmers had adopted these practices by the second planting season.

A cropping calendar that incorporates traditional knowledge on weather and farming practices has also been produced in English, and will be translated into Kiswahili and Kikamba (the national and local languages, respectively). It outlines suitable agricultural and livestock production activities to be undertaken during the rainy and dry seasons. It also highlights the importance of early land preparation, selection of appropriate seed quality and variety, and conservation of livestock feed. The cropping calendar provides guidelines for planting that take into consideration possible rainfall forecast scenarios under different soil types. It provides information on appropriate crop types, seed varieties, planting dates, as well as depth and spacing of seeds. In addition, guidelines on land preparation, application of manure, pest and disease control, weeding, crop rotation, grain selection, packaging, storage and transportation are provided. It aims to enable farmers in Sakai and other areas with similar agro-ecological and livelihood characteristics to make decisions that will result in appropriate farm operations and high production levels. At the time of writing this calendar was being finalized and translated into Kiswahili and Kikamba. The plan is to distribute it to farmers throughout the district (through ALRMP and agricultural extension officers). To date, three brochures that provide information on expected rain and appropriate crop and seed varieties have been produced and distributed in the area.

To enhance perennial availability and accessibility of water, two sand dams, *Kwa Dison* and *Kwa Ndeto*, have been constructed. They are designed to form a partial barrier across a river/stream, which traps sand and water as the river/stream flows. Sand dams are suitable for the area because they conserve water that the community can draw and use domestically and in the farms during dry seasons. There are plans to also drill shallow wells to conserve water. To facilitate this, the project has purchased an auger drill. Training on operation and management of these water conservation infrastructures were provided by the project and the technologies will be supervised by the district water engineer (via ALRMP) together with the community (through a special committee that will oversee management of water conservation infrastructure) once the project leaves the area.

In an endeavour to diversify the economic livelihoods of the community, five selected self-help groups have been strengthened through provision of training on entrepreneurial skills and financial management. The groups, which proved to have clear vision and good leadership, developed business plans outlining their business objectives, and planned entrepreneurial activities, including establishment of tree nurseries, table banking (a short-term, non-formal credit facility that offers low interest rates and has minimal collateral requirements, used widely in rural areas), rope manufacture and trade in kerosene. The micro-credit scheme aims to enhance alternative income generation to complement farming as an income-generating activity.

Resource implications

The project was implemented by six full-time field staff, supported by a further six technical and administrative full-time staff. The number of beneficiaries rose as the project progressed: each farmer who received capacity building and assistance from the project committed to supporting two others through farmer-to-farmer training and the provision of good quality seeds from their farms. The numbers built from an initial group of 40 farmers, to 160 farmers during the second rainy season, to a total of 320 at the time of writing. The total number of beneficiaries is expected to reach 1,280 by the close of the project in June 2009. The benefit of increased water availability as a consequence of sand dam building during the project was felt by other community members not included in the count of direct beneficiaries.

The total budget for the project is approximately US$307,000, of which materials and equipment costs account for around $7,500. This equates to an approximate cost per beneficiary, based on the 1,280 direct beneficiaries anticipated by the close of the project, of $240. Note, however, that this is a provisional figure excluding in-kind contributions made to the project and assuming that there are no further direct beneficiaries.

Lessons and challenges

The concept of climate change is still fairly abstract for many people. In Sakai, there is a general awareness among community members that the weather has been changing, and that these changes are increasingly affecting their lives negatively. However, it is widely perceived that these changes are caused by other factors such as punishment by God or the degradation of natural resources such as forests. They have not yet been linked to the increased emission of greenhouse gases. There is a need for simplified information on the science of climate change and greater awareness creation among local community members, policy makers and other practitioners.

Communities in the area have been adapting to changes in weather and the environment for a long time, and have demonstrated that adaptation to environmental change is possible. With the changing climate, the scale of adaptation efforts has had to increase, making it necessary to augment indigenous coping strategies with modern, science-based knowledge and to introduce modern technologies for adaptation. It is also important to link adaptation to local development in order to encourage active stakeholder participation. To improve the probability of success, consultative and participatory approaches need to be adopted at all stages of the project cycle. This promotes a strong sense of ownership and enables all stakeholders to learn as they work together. In addition, it enables the early identification of challenges and provides room for exploring the most cost effective and efficient ways of addressing them.

The key to effective community-based adaptation to climate change is proper utilization of weather and climate information (which needs to be communicated in a meaningful and timely manner). To effectively communicate the uncertainty of climate, information on possible climate scenarios and their impacts (based on current scientific knowledge) should be communicated. Clear explanations about the uncertainties should be given and stakeholders encouraged to explore a range of potential interventions that could enable them to adapt under different scenarios. Close collaboration with meteorologists should be fostered. Additionally, diversification of livelihood options and demonstration of the tangible benefits of adaptation are crucial. Improvement of agricultural yields, quantity and quality of livestock, access to good quality water, household incomes, environmental quality and human health in changing weather conditions are a clear indication that the community is effectively adapting.

Anecdotal evidence from Sakai shows that agricultural yields have improved since the project began. According to one of the farmers, an initial provision of 2 kg of good quality, drought-resistant maize seeds purchased by the project has yielded a huge change in his livelihood. Using the knowledge and skills gained through training, as well as weather information provided through the project, he subsequently harvested 50 kg of maize at the end of the long rains season in 2007. Out of this harvest, he selected 6 kg of good quality seeds and planted them during the short rains of the same year. Out of this, he harvested 400 kg of maize.

Since the completion of the first sand dam (*Kwa Dison*) in mid-2007, community members have already begun to realize the benefits, such as potable water available in close proximity. This water has so far been used for domestic purposes and the cultivation of kitchen gardens. The farmers have also been recording data on the quantities of inputs and outputs used throughout the year. A follow-up household survey is due to be conducted at the end of the project, allowing comparison with the baseline data and thus clear evidence of project impacts.

To sustain and upscale adaptation projects successfully, it is crucial to involve relevant government ministries and agencies in the project from inception. In the Sakai pilot project, the Arid Lands Resource Management Project, housed within the Ministry of Special Programmes, has been an instrumental partner. Through ALRMP, relevant government representatives have been engaged at divisional and district levels. They have also enabled upscaling of interventions to other arid and semi-arid lands districts. ALRMP will also continue implementing project interventions in Sakai after the current funding phase ends. However, integration of climate change adaptation into national policies has so far been a major challenge. It has been noted that there are capacity constraints among the project team with respect to adoption and application of appropriate tools and methods of policy integration. Plans are under way to conduct relevant training on this including the use and application of tools such as cost–benefit analysis, cost-effectiveness analysis, multi-criteria analysis, and opportunities and risk of climate change and disasters (ORCHID), among others. Also, suitable avenues for integration of climate change adaptation into policy including National Environment Action Plans, Poverty Reduction Strategies, and major development projects and programmes will be explored. Another challenge is the lengthy process of policy review, and exogenous factors that affect policy change such as political interests and prevailing economic priorities. Currently, the Kenya Government acknowledges climate change as a significant challenge to national development. A National Climate Change Office is being set up and will be charged with the task of formulating a national climate change strategy on adaptation and mitigation (where feasible). Lessons on the process of implementing a community-based adaptation project and the impacts of adaptation can and should be drawn from the Sakai pilot project experience, and will be disseminated in the form of policy briefings.

At field level, effective community mobilization was problematic, particularly for construction of the second sand dam. Construction took place during the long rains season and community members preferred to work on their farms (taking part in dam building was seen to distract them from other activities that have tangible benefits). In addition, morale was not very high because of the perception that some community members (who did not plan to participate in the construction) would benefit from other's efforts (free-riding). It is hoped that, having seen the benefits that the community around the *kwa dison* dam are reaping, others will be motivated to participate in these kinds of activity (because they will benefit). Other challenges included the high costs of construction materials, sudden attacks of crops by pests and diseases, and erratic weather.

Conclusion

This project worked with a consultative and participatory approach at all stages of the project cycle, engendering a strong sense of ownership and enabling the project partners to learn together as the project unfolded. The aims of the project, developed collectively by the stakeholders and community members, incorporated three significant elements of adaptation: resilience, vulnerability reduction and policy influencing. Resilience, in the form of crop and seed diversity, was used to address uncertainty in forecasting predictions, while a thorough and participatory baseline survey identified the community to be vulnerable to increasingly frequent drought, prompting a technical innovation in the form of a river dam to address frequent water shortages. Resilience was also fostered through the introduction of seed bulking and storage, enabling the community to secure and control a reliable supply of quality and diverse seeds.

With the involvement of ALRMP, the project was well placed to secure influence over adaptation policy in other parts of the country, ensuring that lessons from the project were connected to a wider network of communities struggling with the challenges of climate change. ALRMP, a government body, was ultimately a key element in the success of the project, enabling activities to be sustained beyond the life of the project and playing a critical role in scaling up the project to the district level. The involvement of ALRMP ensured that the project had built in vertical network connections, with the community, project team and relevant ministry personnel brought together around shared goals. Effective scaling up was also supported by significant efforts before the project began to identify a community that was both representative and willing to be involved.

The involvement of a meteorologist in the project team was significant, and supported the project's engagement with seasonal forecasting to assist farmers in a time of increasing climate variability. Before the project commenced, forecasts were failing to meet the needs of farmers, unable to provide sufficient detail for agricultural planning and management. A combination of training of community members to understand forecasts, provision of local forecasting information, and partnership between the meteorologist and agricultural extension officer ensured that information was provided in terms that were useful to the community, focusing on suitable crop and seed varieties for the coming season. Importantly, while the farmers were unfamiliar with the use of statistics in forecasting, they were prepared to plant a diversity of crops to guard against unexpected weather. However, lack of clarity in medium- and long-term climate change predictions prevented their use in the project – albeit with the intention to introduce them into decision making once uncertainty had reduced. By limiting access to unreliable climate information in this way the project team operated as a necessary intermediary between the community and climate scientists, delaying adaptation activities focused on future climate impacts until those impacts are better understood. In this way the project was able to focus on no-regrets vulnerability reduction and resilience measures that met the short-term needs of the community and secured their interest and involvement in the project.

CHAPTER 7
Multiple pressures on pastoralism in semi-arid Niger

Prepared with Jeff Woodke, Director, JEMED, Niger

Abstract

This chapter describes a long-term project undertaken with the semi-nomadic Tamasheq people of central Niger. Work ongoing since 1990 has sought to build capacity to survive in a drying climate against a backdrop of agricultural encroachment onto the pastoralists' traditional lands. The project employs a strategy that seeks to recognize and build on the strong traditions of the Tamasheq. By providing a fixation point to which the pastoralists can return during the dry season, the approach enables essential community-based development and adaptation activities to be undertaken. In this way the Tamasheq have regenerated degraded land, diversified livelihoods and been exposed to education and health care services. Perhaps most significantly, communities are starting to recognize the need to engage in political processes so as to fight for a policy environment that will allow them to continue to adapt in the face of climate change.

Introduction

Niger is a landlocked country in the semi-arid Sahel region of Africa. Increasing rainfall variability and temperatures and three severe droughts between 1973 and 2005 have had devastating environmental and socio-economic impacts on the former French colony. Nowhere is this more true than in the central pastoral zone of the country. This region, known as the Azawak by its inhabitants, is semi-arid grassland, receiving between 250 and 300 mm of rainfall per year. It is composed of a number of valleys separated by plateaus, which are scattered with fossilized dunes and small depressions. Rainfall has always been capricious in the region. Even in a normal year some areas may not receive any rain at all. Based on their knowledge of rainfall patterns, the Tamasheq people have developed a semi-nomadic or *transhumant* way of life. Normally, a nomadic camp is a group of up to 100 families, usually of a similar family lineage. This larger group is divided up into numerous smaller groups of four or five households. In the rainy season these small groups move together in a northward migration known as *transhumance*. This is usually to the same area, depending on water and grass availability. In the dry season the camp occupies a fairly well-defined area, which they recognize traditionally as their own. They will only leave it if forced to by failing water or pasture resources.

Rainfall amount and patterns have changed considerably in the Azawak. There have been numerous low rainfall years since 1970 and average precipitation in the area has reduced to around 250 mm. Not only is the amount of rainfall changing, but large out of season rainfall events are occurring with greater frequency. The temporal spacing of these events is becoming greater, causing flash floods as well as problems for the life cycle of herbaceous plants. Once broad and forested valleys have now become narrow treeless gullies. This increasing aridity has changed the species composition and diversity of both ligneous (particularly important for camels and goats) and herbaceous plant communities. Less palatable grass species have moved in and numerous more palatable species have disappeared from the area. Tree species numbers have reduced and plants that require more water have disappeared or become very rare, including various types of wild melon (*Citrillus* sp., *aleked* in Tamasheq). Water holding plants were very important to the water balance for humans and animals living in the area. As a food source they provided water and were rich in vitamins and nutrients. The loss of these species has led to an increase in groundwater use among the local population and their animals. It has also led to a decrease in milk production. Milk is the staple food of the pastoralists living in the Azawak and its decline has in turn led to increased water use for drinking and cooking, increased cereal consumption and food insecurity. It has also greatly reduced the mobility of these once nomadic populations.

Having lived as semi-nomadic pastoralists for hundreds of years, the Tamasheq people are tightly bound to their traditions and to the land. In the past, change has been resisted as people did not understand the need for it. However, climate change has been an issue for the *Kel* Tamasheq (*Tuareg*) people of the Azawak region of Niger since a great drought and associated food crisis of 1973. A number of bad years repeatedly decimated the Tamasheq's animal capital after 1973, and following a second major drought in 1984 certain Tamasheq communities realized that these were not simply anomalous occurrences, but that something had definitively changed. Many nomads who had lost their herds moved to towns and villages, and farming seemed the only alternative source of livelihood. Through various projects herders were encouraged to get involved in rain-fed farming or in market gardening. For the most part the Tamasheq people did not want to become sedentary and preferred to try to rebuild their herds, abandoning farming as soon as they had animals again. Others kept farming but this required them to reduce their mobility and change the breed composition of their herds. However after 1984, both the non-farming as well as farming herders realized that modifications to their lifestyle were necessary, and adaptation to climate change came to the Azawak.

Four types of land loss drive the need for adaptation among the pastoralist community: land lost due to desertification; an overall drop in productivity of remaining rangelands due to soil degradation and species disappearance;

encroachment by farmers onto grazing land; and an increase in transitory herders from the south. In the past southern farming villages did not have many herd animals but, as a response to reduced production from croplands, farmers have diversified their incomes and many now herd animals as well as farm. This problem is compounded by a bias within the administration of Niger in favour of agriculture and national policy strongly promotes agriculture and cereal self-sufficiency. There are pastoralists who have seen their southern lands taken over by fields and are also obliged to go north in the growing season. The pastoralists of the Azawak are thus seeing the grazing lands ravaged by climate change, and by encroaching agriculturalists. The resulting tension between the Tamasheq nomads and the government has led to two armed conflicts in the last 17 years.

Agriculture puts tremendous pressure on the resources of the Azawak, especially pasture and surface water. These are treated as common resources and cannot be controlled or managed by the local residents who have no authority to refuse access to surface water and pasture resources. The Tamasheq people's lack of animals and lowered milk production means that they are not as mobile as they once were. Their increased dependence on cereals for food means they have to stay near a market. They do not have the resources to purchase grain in bulk and could not transport it if they did. Even if they could leave, there is nowhere else to go: the pressure for resources is the same over the entire Azawak area.

In spite of these seemingly insurmountable obstacles, the Tamasheq people have resolved to make the adaptation changes necessary in the face of climate change in order to produce a sustainable form of nomadism. Since 1990 Jeunesse En Mission Entraide et Développement (JEMED), funded by Tearfund UK, has worked with the Tamasheq people of the Azawak towards sustainable development of their communities and adaptation to the changing climate. The programme described in this chapter is part of an adaptation and disaster risk reduction strategy which has been evolving over the last 18 years.

Community-based adaptation among the Tamasheq

JEMED's current programme is focused on developing a sustainable form of nomadism based on a fixation point strategy that allows the pastoralists to continue with their rainy season mobility while reducing their need for mobility during the dry season, except in case of crisis, through improved infrastructure and management of water and pasture resources. The strategy aims to strengthen livelihoods through an integrated approach. The fixation point strategy provides for a diverse number of activities to be undertaken in an integrated manner through a community-based approach. In addition to the water and other natural resource elements discussed above, the strategy also addresses food security and livelihoods issues.

The overall goal of this project was to help achieve an environmentally and socially sustainable nomadism in the face of significant climate change.

Key elements in the elaboration and implementation of this strategy are the following:

- *Finding and working with key local individuals.* These people can provide excellent knowledge and perspective as well as invaluable help with local authorities. The majority of staff are from the communities, with a few expatriate staff members.
- *Using participatory rural appraisal (PRA) techniques and ample time* to gather necessary data and develop trusting relationships between the local community and the organization. Follow up needs to be regular and with ample time for relationship building.
- *Having a community-based and driven approach,* where basic ideas are taken from the community and where activities are designed gradually to increase the community's awareness, understanding and ownership.
- *An integrated set of activities.* The effects of climate change have an impact on many areas of society in a least-developed country context.
- *Adequate time and resources.* Successful adaptation will not happen overnight, especially in traditional societies. A normal three-year funding phase is inadequate to make lasting change. JEMED has taken a 20-year approach to the problem.

The strategy has been developing since 1990, when JEMED began working in the area, and is based on a series of fixation points selected by the community. Fixation is not sedentarization. Fixation allows the people to develop the dry season area of their choice, so as to provide the physical and social structures necessary for sustainable development and adaptation. Physical structures include wells, schools, grain banks and environmental regeneration structures (such as dykes) to be built at or near each site in the bush. The social structures include programmes for animal loan, adult education, health care, natural resource management and advocacy. It allows the community to manage the water, forest and pasture resources of their dry season range more intensively, without all the families having to live in one place. It also encourages a modified rainy season migration. The strategy is thus community based and employs an integrated approach that retains mobility while improving water resource management.

The project initially spent some weeks building relationships with the community prior to raising funds for new work. PRA techniques were used to determine the real and perceived needs of the entire community and to gather baseline data to identify community make-up, priorities and livelihood strategies. Socio-economic information data was collected on social make-up, sources of income, family size and demographics, cereal consumption, herd size and composition, water use and availability. Estimations of income were attempted, and where difficult, average household expenditure calculated. Nutritional status was assessed via upper arm circumference measurement and resource access and use was mapped. Finally, discussions were held to identify existing coping mechanisms, the desire for change, and the community's knowledge of possible options for change.

Assessing climate knowledge and awareness

Older people were the source of knowledge of traditional coping strategies and longer term climate trends, as experienced over the past few decades. This information is valuable because it gives a contrast between current climate variability and the climate variability and associated coping strategies of the previous generation. The only modern source of climate and weather information that people in the Azawak can access easily is via short wave radio, where people can get a weather forecast. JEMED/Tearfund carried out a study which gathered the perceptions that local people have of climate change. At all the study sites women and men of all ages were able to give information on the perceived changes in the climate and the impact that these changes had brought. The community members talked of how the weather had changed over the last 20 years and also how the land had changed. The Tamasheq people have a concept which they call *Iban n albaraka* which translates as 'lack of blessing'. According to them, the rain does not fall as it once did. Rainfall is increasingly poorly repartitioned in space and time, diminishing in overall quantity and falling during a shorter period. Water, even if available, never seems to be enough. The grass seems to lack vitamins even in a good year when it grows well. Animals are weak, unproductive and often sick, lacking milk and often miscarrying. Traditional coping strategies could be recited in detail, as well as modifications to those coping strategies necessitated by climate change and variability. These coping strategies related to mobility of people and livestock, and changes to diet.

People's perception of the changes is backed up by scientific evidence. The reduction in rainfall amount and changes in patterns are real, and have reduced vegetative cover, exposing the soil to wind. The wind, along with any run-off water, erodes the soil. The Azawak has also witnessed an increase in wind and dust storms. These storms can blow for days, and over time have blown in sand and dust from the desert. This sand has little nutrient value, but it covers over the more nutrient-rich soil beneath: even if pasture germinates and grows it cannot access the nutrient rich layer deeper down. The result is a pasture which lacks calcium and phosphorus, two essential ingredients for milk and animal production.

In some areas of Niger, there has been success with reforestation, and a re-greening process is under way. However this is in a small area of the country, where rainfall is still above 300 mm and rain-fed agriculture is (marginally) profitable. Satellite images appear to show improving vegetation cover in the Azawak as well, but on the ground the species composition has changed and many of the plants are unpalatable for livestock.

Social networks

Building local social capital and networks begins with meetings with the site leader or person of influence. This person is then asked to help JEMED organize community meetings. Participation varies; at the right time of year close to 100 per cent of the men will participate and 20 to 25 per cent of the women. After an initial set of group meetings tent-to-tent visits are made to meet with individual households. Typically by the end of this stage of activity the community will have elected a management committee, have determined the problems the committee should address, and possibly the technologies to use. This community ownership and involvement is crucial for success of climate change adaptation, yet acting as a community (not just looking out for individual interests) is not easy for the Tamasheq people. This is the area where most problems occur. No incentives are offered for individuals to get involved other than the expected benefits of the project activities themselves, which will benefit individuals as well as the community. If the job of animation/mobilization has been well done, people will be motivated to work with the project. Failure has usually occurred where insufficient time has been invested, or where the leadership is self-centred.

The Tamasheq people have a deep distrust of politics and prefer to let their chiefs deal with political issues. Lack of education and political participation has hindered development and democracy in the Azawak, but this fear of politics and education is changing. The residents are getting a clearer idea of what democracy is and participation is on the rise. There is now a major opportunity to develop a more active civil society, building the capacity of local NGOs and CBOs which exist as informal associations to play an active role in the development process of the Azawak region. In particular, the current process of changing the rural code laws of Niger is an important opportunity for the communities. There will be a new pastoral code for land in the pastoral zone. In general, the proposed pastoral code strongly favours mobility as the most economic way of herding in the pastoral zone of Niger. However, this approach does not reflect the increasing lack of mobility of herders due in large part to climate change. It also does not favour fixation, or land and resource management by local communities. The new code is therefore not conducive to pastoralists' education, food security or improved health care, as these services require some fixed points for delivery.

In early 2008 JEMED funded a workshop in which all five of the communes in the Abalak department came together to make suggestions to the government for changes to the text of the proposed pastoral code. Suggestions included: giving priority use rights to local communities; allowing the process of obtaining priority use rights to be controlled by local elected officials; including fixation as a viable method of herding, while enabling mobility for those who wish it; and allowing access to transitory herders. Clear concise laws would allow for a reduction in the tensions which currently exist between residents and non-resident herders in many parts of the pastoral zone. The document which was created was sent to national level officials in Niamey in charge of the development of the pastoral code. The local communities will be encouraged to push the issue with parliamentarians.

Social networks were also developed in the course of technical innovations. In seeking to establish rotational grazing patterns some community members sought to cooperate with neighbouring groups to collectively manage a larger area. This raised the issue of land tenure, which the pastoralists of the Azawak do not officially possess. However, a number of community leaders came to fully understand the political, environmental and demographic issues associated with land tenure and adaptation. They are now actively involved in advocacy work, fighting for land rights at the local and national level.

Awareness raising on climate change

No specific programme was geared towards raising awareness on the causes of climate change, but information about climate change was shared regularly with people and prefaced a number of activities. Communities in Niger are well aware of the impacts of climate change and project staff raised the level of understanding of people about the process, explaining how lack of rain has an effect on pasture production and land, and taking a more regional perspective in place of the local view that pastoralists were accustomed to taking. They had tended to look at pasture production locally and determine only if it was sufficient for their needs. They did not consider the impact of transitory herders or the availability of pasture in adjacent areas. This broader perspective requires a capacity for abstract thought which is not easy for people lacking formal education. By increasing understanding of the processes and impacts of climate change (although not of the science behind the causes), the communities were better equipped to be proactive and make decisions.

Technology development

Improving water resource management

Before the 1960s, few deep wells existed in the Azawak. The people used seasonal ponds which are open to all, until they dried up. At this point they would move to areas where they could dig shallow temporary dry season wells. In areas where there was no shallow water available, some traditional deep wells existed. In the 1960s the government began putting in pump stations, which opened the range up to many non-resident herders. This caused overgrazing around the stations and also encouraged overstocking and inefficient herd management, culminating in disaster in 1973 when large numbers of stock were lost.

Subsequently, deeper, hand dug wells emerged as a dry season water source. These wells were essentially private, owned by the person or community who dug them. Thus the owners can restrict who gets water and to some extent control the nearby grazing. Rather than relying on pumps, animal traction was used to draw water, allowing the wells to be dispersed over the range and better utilizing the dry season pasture.

The initial part of creating a fixation point was to secure a source of drinking water by digging a deep well, owned and managed by a particular community. Local techniques and ideas were refined to develop an improved, traditional well-lining method, allowing a safe and secure well to be dug for one-third of the cost of putting in a concrete lined well (see Box 7.1).

Improved natural resource management

The fixation point strategy aims to enable a more intensive use and management of dry season pasture, forest and water resources, which is key to successful adaptation when climate change and population pressures together are leading to a decline in available natural resources.

Initially the people saw the environment as the responsibility of God alone. For example, when asked about de- and re-forestation, they said that God gave the resources for people to use and that if he wanted a new tree, he would make it grow. Yet all Tamasheq understand fully that their lives are tied to the grass, trees and water: in their hearts they resisted change, but in their heads they knew that it was necessary.

The fixation point strategy allows for incremental steps in resource regeneration and management. It begins with passive techniques for regeneration and improvement, such as loose rock dykes or contour

---

Box 7.1 Well construction technology

Traditionally, cement did not exist in the Azawak. This made deep wells almost impossible, except in areas where the substrate was solid all the way down to the water table. In the first few metres of such a deep well, wood and mud would be used to reinforce the unstable soil and subsoil above the consolidated substrate. Still, few deep wells were dug until the 1980s.

With the advent of cement the traditional technique has also changed. Now wells could be put in areas which have less than stable substrates. If a layer of sand or clay occurs along the well depth, it is excavated laterally to about 80 cm. This space is filled with rock and cement, and then covered with plaster. This technique is not suitable for layers more than 6 m in thickness as the lining becomes unstable. To get around this, two methods are used: one is cement bricks and the other is to use the rock lining, but with anchors made of reinforced concrete which goes deep into the substrate laterally every few metres. This supports the weight of the rocks and cement above. The cement brick method also uses the anchoring technique.

This method is suitable in all but the most unstable substrates and has been used to sink wells up to 145 m deep. This type of well is cheaper than a concrete-lined well because the lining is only used where needed. It is also cheaper by the metre than a reinforced concrete ring and easier to repair. If a concrete ring slips or cracks and falls, it can cause immense damage which is often irreparable. If a section of improved traditional lining gives out, it is easier to replace and repair.

bunds which, after initial construction (involving labour from all the male population), require little annual maintenance and allow regeneration to occur with little or no input from the population (Figure 7.1).

Dyke length varies from 50 to 300 m depending on the area to be treated. More than 24 of these dykes have been built, each one regenerating an estimated 25 ha minimum and serving a population of at least 70 households; some dykes are having a beneficial impact up to 1 km upstream. The regenerated land is sometimes protected by a fence but enclosures are limited to 400 m square so that the mobility of transitory herders is not affected. The dykes give a visible impact and tangible benefits: after one rainy season they were able to harvest wild wheat from the site for the first time in 20 years and older people identified herbaceous plants that had not been present for the same amount of time. Young trees began to sprout. These highly visible results were key to developing understanding among the community that their efforts could not only stop but also reverse the effects of the changing climate on their local resources. The dykes were thus an excellent way to begin the process of land management that is essential for adaptation. Once a community gains understanding, it is easier for them to progress to more active management techniques such as exclusions or rotational grazing plans.

Improving food security

In terms of livelihoods, the main emphasis of the project has been to strengthen the principal income generating activity (livestock) through animal loan programmes. The project also helps people diversify into other income generating activities, primarily focusing on women who are otherwise not involved in income generation in Tamasheq. Women selected by the women's management committee have been supported to set up shops, but the profits

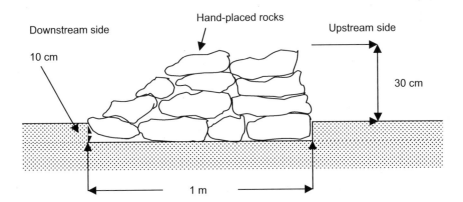

Figure 7.1 Cross-sectional view of a loose rock filtration dyke; water flowing from upstream will back up behind the dyke and slowly filter through the rocks
Source: J. Woodke, JEMED, Niger

are small and growth is slow. Most people would like to access micro-credit and revolving credit funds to develop small trading businesses.

The Tamasheq have never seen availability of food as a primary indicator of food security: access to a market and adequate purchasing power to access food available in the market are the critical factors. To improve access to food the construction of grain and fodder banks has been promoted, as well as the harvest of natural grains (wild wheat) on regenerated lands (integrated with natural resource management efforts). Formerly, without a fixation point, there was nowhere for these activities to be located. All the storage buildings are of mud brick construction. The bricks are made by the local population and a mason is contracted to construct the building. The community is responsible for a number of activities during construction and for sourcing some materials. The project pays for the mason, cement doors and windows, any anti-pest treatment desired and some hardware.

A programme has been initiated to develop the communal maintenance of the buildings. A herd of animals is loaned to the community and after two years the proceeds are used to maintain the infrastructure. This has now been implemented at four sites where it is working well. Two more sites are scheduled for this year. However, initiating this process takes a high degree of trust between the project and the community, and a high degree of communal commitment is necessary to make this programme work.

Resource implications

JEMED's programme staff currently number 11 (including the director) of whom seven are field staff. For the integrated programme it takes between 5 and 7 years for a site to reach an acceptable level of sustainability, with complete sustainability after a total of around 10 years. The project has reached 1,890 families, each with an average of eight members, yielding an overall cost per household of around £1,890 (US$2,820) over the 10 year period, or £23 per beneficiary per year.

About 10 per cent of the total budget is allocated for specific training (rising to 13 per cent if transport costs are included), while 90 per cent of the activity budget is for hard resources such as materials and tools.

Lessons and challenges

The effectiveness of the project interventions was particularly evident in 2003 and 2004. In 2003 the project area suffered heavily from locusts, a year before a major plague in Western Africa. There was virtually no ligneous foliage available for livestock in the Azawak. The localized catastrophe was not reported in the news and no aid was sent. However, people from the project sites were able to maintain the mobility of the herds, while the women and children remained at the fixation points. In 2004 a severe drought struck the region. The project communities and JEMED began reacting in September

2004, about seven months before any declaration of crisis by the international community. People from different project sites worked together to survey pasture in the entire region, establishing the areas where grass would last the longest. De-stocking took place and animals were moved before the end of the year. Extra stocks of grain were purchased and pasture from behind water harvesting structures was reserved for draught and milk animals so that the women and children could stay at the fixation points. When the food crisis was declared in March 2005, these long-term sites were already secure and required minimal relief aid.

JEMED attributes the success of the programme to a number of factors, discussed below.

Shared faith

JEMED staff are firmly committed Muslims or Christians, creating a level of commitment from staff to the people as well as the activities. They have an attitude of service to God and man which encourages values such as integrity and honesty. The beneficiaries are also religious for the most part. They appreciate the faith of the project staff and this creates an atmosphere of trust, confidence and acceptance.

Finding the right people

In 1990 JEMED met with a local Tamasheq leader in the Abalak area. At that time he was the only one in his clan to have ever received any kind of formal education. While most of the Tamasheq people knew that something was changing in the climate, he was one of the few that realized the change was permanent, and that the people must adapt to it. Finding this key person, who had an understanding of the problem, as well as holding a position of influence, was of fundamental importance to JEMED efforts. Based on his knowledge of the culture and his understanding of the problem, he had already developed a framework of ideas for community development and adaptation. He was also able to provide JEMED with access to the nomadic communities which would otherwise have been impossible.

Employing community facilitators

People from the communities were employed as extension workers or community animators/mobilizers. This enabled in-depth access to the community and built a high degree of ownership. One challenge is that work proceeds slowly because of the often low level of education. However, this approach has the great benefit of building the capacity of the community because these individuals gain an understanding of the issues. Two former employees left to start NGOs of their own and three others have been elected as mayors of their local communes, thus adding very significantly to social capital.

Slow and steady

While the full programme contains a number of activities, they do not all happen at once. Adequate time was seen as crucial to adaptation work. The activities require different levels of community involvement and ownership. In 1990, dyke building was the first activity. It showed immediate results in terms of increased grass growth above the dyke, and gave the people physical evidence of regeneration in a relatively short period of time. Once an activity showed a positive outcome, the project staff were able to build the community's capacity to move on to the next activity. At sites where project personnel were physically chased away in 1990 when the programme began, there are now schools which have produced high-achieving students and individuals serving as mayors and communal council members.

Increased climate change awareness

Increased understanding and awareness about the changing climate enables communities to become proactive in disaster management. In 2004 after the rainy season was over, many of the project sites were able to understand that a crisis was coming; this was in stark contrast to most other people in the Azawak who felt that somehow things would recover. There was proactive de-stocking, in which people took their animals to southern markets where they got an excellent price. They were able to purchase grain in the south while it was still available. This was a first: pastoralists in the area had never previously voluntarily accepted de-stocking.

Staying small and focused

JEMED has resisted the temptation to run a larger programme. This has allowed focus to be retained on a limited number of sites and enabled sufficient time to be spent to see change effected. It also allowed the organization, because of its low budget, to operate without the scrutiny of the government, functioning more as a local NGO and community organization than an international NGO. Rapid growth in response to drought crises, when the number of project sites grew from 18 sites to 45, caused problems: budgets rapidly grew very large. The balance of staff is not yet correct: budgets show a low percentage of personnel and running costs with most funds spent on activities. While this finds favour with donors, it makes it hard to have the necessary level of relationship-building at new sites.

Assessing programme impact

Excessive reliance on quantitative impact indicators (for example, increasing household income by x per cent) can lead to undervaluing some outcomes that are valued by the beneficiaries. One example is that the benefits of land regeneration are not adequately represented by the number of hectares restored, since it is difficult to know in advance to what use the regenerated land will be put. Nor does the area regenerated give the true impact of the intervention on local livelihoods. In one site, this might be represented by the increase of wild wheat production, while in another, women are able to have pasture for the donkeys in the dry season when the men are gone to camps. The latter means that the donkeys can continue to draw water and so people can stay at the fixation point and the school can stay open. It is important therefore to undertake a participatory evaluation of project impact.

In late 2005 JEMED conducted a study to look at the impacts of its disaster relief efforts, and to determine what was needed for post-crisis recovery. Over 600 households in three distinct groups were surveyed in three groups. First were long-term development sites. The second group contained sites where crisis relief had been provided. The third were control sites, which had received no relief aid and did not work with JEMED.

The results showed that the long-term sites had an average of 20 per cent less drought-related stock loss than the crisis relief sites (the crisis relief also included supplemental animal feed), and an average of 40 per cent less stock loss than the control sites. When only large stock (cows and camels) were considered, there was up to 70 per cent less stock loss. The long-term sites had no drought-related fatalities and no cases of malnourished children. There was no drought-related exodus to urban centres and schools stayed open. Many of the sites had actually become places of refuge for communities which had not received aid. Because people had retained a significant portion of their assets, a return to normal life was attained quickly. This is in contrast to many other places which are still suffering from the effects of 2004–5.

Another study completed in December 2007 shows that malnutrition is much lower, and general health and maternal health much better at the long-term sites (compared with crisis sites or control sites). As the programme is aimed at promoting a long-term approach – adaptation to climate change – the beneficiaries have been able to react to and recover from severe drought and food crisis with minimal loss of assets. Despite drought and continued climate-related environmental and social degradation, they are able to maintain their chosen lifestyle and economic activity, and attain a measure of prosperity.

Improving the programme

One technology that failed, but that has worked in similarly arid conditions, is tree planting within half-moon catchments; the failure was due to the windy conditions of the area and the lack of an inexpensive way to protect the planted trees from the wind. This highlights the importance of understanding the local context and environmental conditions in selecting appropriate technologies.

Education is crucial for long-term development and adaptation. It allows people to think abstractly which greatly helps them to understand the issues associated with climate change and to generate solutions. The initial literacy campaigns were targeted at all adults. They worked well for three years, but then failed. This was in part because people reached a literate level and saw that their increased skill could not be used because of a lack of materials to read. It was therefore perceived not to be worth the daily time investment (for a three to four month period) to learn more, especially for women. Classes were abandoned and potential new recruits were reluctant to come to a class that seemed unpopular. The approach was subsequently changed to target people who would use literacy skills regularly, such as those working with CBOs who needed to keep records. Another failure was an attempt to introduce woodless construction in order to reduce pressure on the forest resources. It was seen as expensive, and at the time it was introduced, few Tamasheq people were constructing permanent homes, and still preferred tents.

There is also a strong demand for micro-credit and revolving credit funds for individuals to access. These ideas will be incorporated into a future funding proposal.

The importance of an enabling policy and institutional framework

The current rural code laws require that a community secures priority use rights before any management of grazing lands can be done. This process is convoluted and difficult, and the priority use rights are weak. While the community can control the grazing of its own animals through its own rules, there is a strong risk that regenerated pasture, if unprotected by enclosure, will be the first to be consumed by transitory herders (much to the chagrin of the resident population). For this reason, changes are needed to the pastoral code as currently proposed, to provide support for community-led natural resource management. Despite the threat, people still favour building more dykes as they realize that the benefit is long term and can see the regeneration of tree cover, on which the transitory herders do not really have an impact.

Another example of the importance of enabling policy environments is a programme to train and equip paravets (self-employed community-level livestock experts who have received basic training on animal health). This venture initially failed because the government would not allow paravets to vaccinate. Government policy has since changed; paravets are now being trained to have limited vaccination capability and the programme is having some success.

Scaling up

Because of past success JEMED is under some pressure to work with more communities, beyond its own capacity. While this has forced some changes to the nature of its work, (such as spending less time at project sites, and moving more rapidly to new project sites), it has allowed JEMED to scale up activities (on a local level) and to have a greater voice nationally and even internationally.

In terms of scaling up, the following technologies and approaches all are strong candidates:

- land management techniques, dykes and improved deep wells;
- animal loans to enable herd restocking (currently the crisis response plan for Niger does not call for any restocking after a moderate crisis such as that of 2005; this policy gap means that a large population is left highly vulnerable);
- grain and fodder banks to improve human and animal food security;
- revolving credit schemes to improve income diversification and help avoid the necessity for men to migrate in search of work.

Conclusion

Key to the sustainability of adaptation efforts in the Sahel is improved and local control over land and water management. Training for advocacy for land tenure rights is thus a vital component of this adaptation programme. However, donors often favour activities that show benefits within the short term while simultaneously having long-term value. Progress on the land tenure issue is likely to be slow, and does not offer the kinds of immediate benefit that donors like to see, making funding a challenge. One option would be for donors to ensure the issue is built into a national development plan or a poverty reduction strategy plan.

In this programme, the catalyst for adaptation activities was the pressure the Tamasheq were facing on pasture and land resources due to apparent changes in climate overlain on social factors: population growth and movement. Local leaders recognized the need to change some aspects of the traditional way of life to adapt to the new challenges. Central to success of the programme has been that activities have been aligned with the local culture and norms: recognizing the need for mobility, and drawing on extensive knowledge of management of pasture and tree species

The gathering of information during the baseline research built trust and understanding between the programme team and the communities. Collecting local knowledge of changing climate enabled the team to raise awareness about the wider implications of these changes: a permanent change in climate, and increased population pressures, together with a rural land code that does not support the traditional coping strategies.

In developing technologies, availability of local resources and affordability were key criteria, with the focus on improving existing locally used technologies. In a traditional society where social norms and livelihood strategies have been developed and adhered to for centuries, implementing change needed to be taken at a slow pace, so that people feel confident and competent. For the Tamasheq, adaptation to a long-term trend of erratic rains and decline of resources demands significant lifestyle change in adopting a settled pattern (fixation) for part of the year (dry season). This change was facilitated because of the clear livelihood benefits of access to education and health, and emergency relief when sudden shocks in the form of locusts or extreme drought occurred.

For the Tamasheq, lack of access to wider social networks is a major challenge for securing the kinds of policy change on land tenure that will be necessary to ensure sustainability of the adaptation strategies adopted. In order to influence policy change the community must take charge and enter the political process; this in turn requires a certain level of education and engagement in the political process. Activities to increase engagement in wider social networks cannot be introduced early on in an adaptation programme, until a certain level of literacy and understanding among a sufficient number of community members has been achieved. The lack of access to government in remote areas means project skills-building activities are particularly crucial. The challenge for sustainability of the programme and for the continued adaptation of the Tamasheq is how to keep people updated particularly on climate predictions and seasonal forecasts. The question posed is, are there knowledge networks into which pastoralist communities can be linked – not only in Niger, but more generally? If so, how can this be achieved using modern technologies that could be applied in these remote areas?

CHAPTER 8
Declining water resources in Sudan's Red Sea coastal belt

Prepared with Dr Balgis Osman-Elasha, Senior Researcher, Climate Change Unit, Higher Council for Environment and Natural Resources, Sudan

Abstract

This chapter draws on a review of a project undertaken between 1993 and 2002 with an agro-pastoralist community in the coastal plain north of Port Sudan. The project sought to strengthen the resilience of a community living in a highly drought-prone region and, while not intended as an adaptation project, provides an opportunity to examine the impact on adaptive capacity. Through the introduction of improved water harvesting and food crop production practices the project was able to increase livelihood resilience, while links with government departments contributed to the sustainability of the project, and the involvement of meteorological staff was critical in informing decision making. However, the review highlights the importance of an enabling policy environment, noting that, while climate change will play a significant role in the future of livelihoods in the area, so too will decisions on the future of water resource management in the region.

Introduction

The material for this chapter has a somewhat different origin from that of the other case studies. The project on which this case study is based was implemented by the NGO SOS Sahel between 1993 and 1997; after that period, activities continued through a partnership agreement between SOS Sahel and the Ministry of Agriculture until 2002. The project – the Khor Arba'at Rehabilitation Programme (KARP) – was undertaken in the Red Sea State, one of 26 states of Sudan, and focused on the community of agro-pastoralists living in a narrow coastal plain north of Port Sudan near the Red Sea. Because the project was regarded as successful in increasing the resilience of its target group, it was selected as a case study for a project under the regional programme of the Assessments of Impacts and Adaptation to Climate Change (AIACC), a global initiative developed in collaboration with the Intergovernmental Panel on Climate Change and funded by the Global Environment Facility.

The purpose of the AIACC programme is to advance scientific understanding of climate change vulnerabilities and adaptation options in developing countries and to generate and communicate information useful for adaptation planning and action. AIACC provided financial support to 24 regional study teams to conduct three-year investigations of climate change impacts, adaptation and vulnerability in 46 developing countries. The case study in Sudan aimed to contribute to the limited knowledge base of possible adaptation strategies for Africa, and had as its objective the investigation of the relationship between the livelihoods measures adopted under the KARP project and their appropriateness in terms of sustainability in contributing to community resilience to anticipated climate change in the area. It also examined means for scaling up successful measures through integration into national development planning. The methodology of the case study included a literature review, group discussions with members of the community, meetings with government stakeholders and NGOs and, most importantly, a detailed interview survey. This was undertaken with a randomly generated sample of households from 5 of the project's 10 sites, in all 47 households out of a total of about 600, around 90 per cent of whom benefited from the project either through direct inputs or with improved access to water and extension services.

The Khor Arba'at region

The region where the project was implemented is the home of the Beja pastoralist and agro-pastoralist tribal groups. It is generally characterized by relative isolation and harsh terrain, highly variable rainfall system with recurrent spells of drought, small area of cultivable land and low population density. Rainfall averages recorded for the period 1900 to 1980 range between 26 mm and 64 mm per annum, most rain occurring as thunderstorms. Much of the rainfall runs off to the *khor* (small streams), and there is also a high loss from evaporation. A combination of factors – the hilly topography, the rocky nature of soils, the intense but sparse rainfall in the area and the poor vegetation cover – means that surface run-off is the only source of fresh water in the Red Sea area.

The project area was the catchment for the Khor Arba'at, covering an area of 4,750 km^2 lying some 20 km to the north of Port Sudan, on the coastal plain of the Red Sea. These lands, long known as *Arba'at Zira'a* (*Zira'a* meaning farming in Arabic), lie within the semi-desert acacia scrub zone, but the specific project area lies on the alluvial deltaic fan of Khor Arba'at. Spate irrigation farming had been practised for many years by pastoralists, taking advantage of occasional floods. Currently only around 1,000 ha are under cultivation during the rainy seasons; a larger area has the potential for being brought into the spate irrigation system, if more water were available. In recent years, agricultural production has focused on vegetable production – watermelon, cucumber, cantaloupe, okra and tomato – for the market in Port Sudan. As a tradition-bound society, land is a highly prized asset among the Beja tribes of the Red Sea hills and its ownership and/or transfer is primarily controlled by the tribal leaders. Unusually, the land in Arba'at has been registered under the names of families since the early 1960s instead of being under communal ownership as is common among the other largely pastoral Beja communities. That early move by the community and local leaders to register land reflects the community's awareness about the high value and fragility of their natural resource base.

The unreliability of rainfall raises the value of Khor Arba'at to the local community and their dependence on it for both drinking and irrigation purposes. The Khor Arba'at is the largest and most important source of fresh water in the whole of the state. Since the 1920s the state authorities have used the Khor to supply fresh water to Port Sudan town and to support agriculture on the fertile land of Arba'at. Over 60 per cent of the state's population live in Port Sudan and depend on the Khor Arba'at for their drinking water supply. Though seasonal in nature and small in terms of volume, the Khor Arba'at supply has, except during very extreme and lengthy drought periods, been sustained as a year-round supply. This has been achieved through the damming of the Khor at its entrance to Arba'at at the foot of the hills, creating a reservoir and controlling distribution through a piped network. However, the fluctuating flow has meant that the community has lived with flows both too low to support farming, and severe flooding.

Historically, the Beja pastoralists – the major tribe in Arba'at – developed or adopted various mechanisms that preserved their environment and allowed for recovery after extreme climate events, both droughts and floods. They adopted a primarily subsistent agro-pastoral system which is well adapted to the harsh condition of the region, selecting animal types which are adapted to arid conditions – mainly goats and camels. They adopted a dispersed pattern of settlements to maintain land carrying capacity, reduce competition and conflict, and allow for population increase. They practise geographical and temporal migration up and down the hills in pursuit of water, pasture and cultivable lands. The larger *khors* represent the main winter and summer resources for livestock and some *wadi* deltas have developed over time into common areas, providing refuge to all groups at times of crisis. Although distance, duration and direction of migration might have changed as a form of adaptation in response to changing conditions, the overall pattern has remained largely the same in most rural areas. The Beja apply a strong system of social sanctions on the use of resources, ensuring pasture and tree protection in particular, which is enforced by tribal leaders. After each drought cycle, the animal stock levels are rebuilt through the Beja tradition (known as *salif*) for managing animal and land resources – the two components of Beja agro-pastoralism. For the Beja, animals represent the main means of economic and social mobility, recognition and survival.

Frequent occurrences of drought and famine conditions in the Red Sea hills have largely been the norm during the 20th century. That traditional pattern of natural short-term recovery was shattered after the long drought and famine of the 1980s and the subsequent failure of the system to reconfigure. The recurrence of drought and famine conditions has made the Red Sea State heavily dependent on central government and foreign aid organization support, and made long-term planning, including that of combating drought, a low priority. The area has almost constantly been in a state of emergency and relief operations only vary in scale, length and location from one year to the other.

This chapter, while not examining an intervention that was designed as an adaptation project, provides an opportunity to look at the extent to which a project designed to strengthen the resilience of a community living in a highly drought-prone region was actually successful in increasing adaptive capacity. This case study is also important as it contributes to understanding about adaptation in a least developed African country and a region in which there is a paucity of adaptation projects. The structure of the remainder of this chapter is as follows. The next section describes the activities of the original SOS Sahel KARP project, while the lessons learnt and conclusions are provided in the remaining two sections and are drawn from the case study research undertaken under the AIACC programme.

Adapting to drought in Arba'at

The Khor Arba'at Rehabilitation Project (KARP) was conceived following a regional investigation by SOS Sahel of potential agricultural development projects in the area. Khor Arba'at delta was chosen as an area where it was felt that rehabilitation of the somewhat degraded farming system could provide considerable benefits to the local community in response to the Sahelian drought in the 1980s. Modern agricultural and development forces in this region had persistently neglected traditional knowledge, so that rural people had become increasingly dependent on ready-made relief prescriptions and development projects. The focus of the state agricultural department has been on subsistence farming in semi-arid areas, and had overlooked the particular needs of the farmers working the more fertile soil in Khor Arba'at. As a consequence they started to lose their capacity to manage their subsistence livelihood systems. The people of Arba'at were eager to participate in a broad-ranging programme designed to improve their livelihoods. The main objectives of the KARP project were to:

- contribute to the rehabilitation of Khor Arba'at delta in order to realize the agricultural potential for the benefit of tribal groups in the area;
- improve livelihoods through sustainable management of natural resources to achieve sustainable food security;
- plan and execute an equitable water harvesting scheme;

- enhance grass-roots participation and the overall development of the community;
- plan and execute a water harvesting scheme by constructing bunds;
- build an extension centre to train the local community and the coordinating committees;
- produce a replicable model if successful.

Required inputs were agreed upon between SOS Sahel and the appropriate state government departments in a co-financing formula with a joint steering committee to plan and oversee implementation and monitor the project progress. Initial activities included conducting a baseline survey and setting up a coordinating committee for development activities in Khor Arba'at.

The partners developed the project activities through a process of shared thinking focused on the following questions:

- How can the water of Khor Arba'at be used for agricultural purposes in a more efficient manner?
- How can agriculture be modernized in such a rural area?
- How can use be made of all available water with its different levels of salinity?
- How can the environment be preserved and kept intact while getting rid of the undesirable tree species such as *mesquite*?
- How can team working among the concerned institutions be initiated and strengthened?
- How can a much more sustainable agricultural production system be achieved in the Arba'at area?

Emphasis was laid on agricultural extension and community training for the acquisition and development of skills, resource management experience and use of appropriate techniques and technologies. The main areas of intervention and activities comprised: water management and harvesting; soil reclamation; training in agricultural practices; community organization; support for education services; and literacy classes.

The intervention also initiated activities solely for the women in the community, including: training in home economics, environmental health and hygiene, childcare, and literacy classes for young and older women who missed the chance of formal education.

Local knowledge of climate variability

Since the project was not focused on climate change adaptation, it did not explore local knowledge in relation to predicting climate events such as the onset of rains or signs of a long drought, but instead focused more on the local knowledge and coping strategies developed and employed by the community through careful observation and understanding of their environment and its meagre natural resources under a variable climate. The KARP project reviewed how local people in the region conceptualize their interactions and maintain their livelihoods through their local knowledge and traditional practices, which include their dispersed settlement pattern and seasonal migration following water and pasture and cultivable lands.

Social networks

The community lives in scattered homesteads along Khor Arba'at. The lack of government support in terms of extension had led to a strong community spirit and the development of their own resources. After the droughts of the 1980s, the local community approached the state government to assist them in harvesting more water in order to expand their cultivable land, both to recover their livelihood and to meet the growing demand for vegetables in Port Sudan. Some earth-moving machinery was brought to the area and earth embankments were constructed to harvest the Khor waters. The community formed a committee for the distribution of the Khor water among the nine sub-villages. The control of Khor Arba'at water has added to its value to the community by permitting utilization for longer periods, enriching the surface water and underground aquifer and allowing vegetation growth, all helping the community to enjoy a settled and stable life, unique to all rural communities of the central and northern parts of the Red Sea region.

The KARP project promoted the participation of women in public life via their inclusion in the production process (they constitute almost two-thirds of farmers) and their involvement in literacy classes and training programmes. The project led to women's involvement in committees rising to four times its level prior to the intervention. Through the project, a farmers' committee was formed, building on a farmers' union that already existed but whose role was unclear to most of the community.

The research study found in its interviews that the overall impact of the knowledge and skills acquired through the relationships made possible by SOS Sahel interventions had been strongly positive. The communities' access to information has substantially improved as a result of the linkages established with government departments by the process of project implementation, their direct and frequent contact with urban markets, and their membership in the State Farmer's Union. In particular, their ability to predict environmental change has improved through understanding of and access to meteorological information from the Meteorological Department and the extension officers. As a result, household capacity to sustain itself during short drought years is currently much higher than it was before. A major benefit to the farmers was gained by obtaining licences for local communities from Arba'at to sell vegetables directly to consumers in Port Sudan's vegetable market. This has enabled people to gain direct access to the market instead of selling to auctioneers and hence has increased the farmers' returns. Marketing their produce allows the community members to purchase their staple crop all the year around. In this way, a cash income is as effective for food security as people having their own grain stores.

Technology development

The project focused on maximization of the effective use of the Khor's water and increasing the productivity of land through a variety of measures. Principal among these was the construction of earth contour bunds and a terracing system to maximize the volume of water harvested and increase the area irrigated (Figure 8.1). The first stage was the strengthening of old earth bunds, followed by construction of additional bunds. Contour bunds are a simplified form of micro-catchment. Construction can be mechanized and the technique is therefore suitable for implementation on a larger scale. As its name indicates, the bunds follow the contour, at close spacing, and by provision of small earth ties the system is divided into individual micro-catchments. The main advantage of contour bunds is their suitability for the cultivation of crops or fodder between the bunds. As with other forms of micro-catchment water harvesting techniques, the yield of run-off is high and, when designed correctly, there is no loss of run-off out of the system. The bunds also increase the recharge of the surface and underground reservoir and, when vegetation cover increases, reduce evaporation.

Other technology developments included: 1) enhanced equity in access to water via a system of piped water distribution and community management of the network; 2) the introduction of improved hybrid seeds for new vegetable crops, and pest and disease control measures (using chemicals); and 3) the introduction of the date palm and home gardens to the area. The latter contributed to food security, increased cash incomes and, with low water requirements and requiring little space, were compatible with the local conditions.

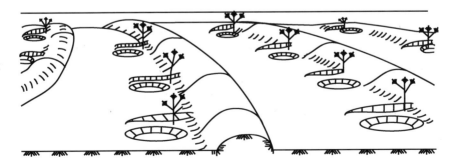

Figure 8.1 Contour bunds for trees

Other interventions

The project offered a credit facility and a large majority of people took advantage of this. It was the community's first opportunity to access credit. The credit system for agricultural inputs supplied by the project was taken over by the farmers' committee; after the project, farmers were dealing directly with the Agricultural Bank of Sudan. The repayment rate reported in a survey at the close of the project was high, comparing well with a much more established scheme in Port Sudan.

The baseline survey revealed that almost all the community members had never received any extension services prior to SOS's arrival in the area. The project provided extension training and advice, including evening sessions, aiming to reach all project households. One of the new opportunities provided by the project was adult literacy classes. Hardly any adults had attended literacy classes prior to the SOS project. That figure rose to almost three-quarters during the project, and for women, the figure reached 70 per cent. High attendance was achievable because the location of classes was distributed widely throughout the project area, making classes accessible.

The standard of education, which was widely regarded as inadequate before the project, is now seen as adequate by three-quarters of the respondents. Although views differed on the proportion of children enrolled in school after the project, the majority considered that at least half of all children were enrolled. For girls' education, the change was even more marked, with two-thirds claiming that all or most school-age girls were enrolled compared with a tiny proportion prior to the project. The contribution of the project to education was direct in terms of improving the school environment (construction), contributing to teaching materials and improving household incomes, thus reducing the families' need for children's labour.

Resources

The project was funded through an unusual development model of a close partnership between SOS Sahel and government institutions. While the NGO contributed some funds, they were more than matched by government institutions, principally the Ministry of Agriculture, and also the horticulture department. It is therefore difficult in this case to provide cost figures comparable to those provided in other chapters. In line with the cost-sharing arrangements agreed, SOS Sahel provided a large truck, costs of transport and field allowances for the ministry-seconded staff and contributed to the purchase of improved seeds; the ministry provided the technical staff and the major contribution to the cost of the inputs, while the farmers started by paying 25 per cent of the market value of the improved seeds, a contribution that was gradually raised to full cost. The Agricultural Bank provided credit for seeds and small loans for interested farmers. Overall, the cost sharing of inputs was as follows: KARP 40 per cent, government 40 per cent and beneficiaries 20 per cent.

Lessons and challenges

This section is based on the findings of the AIACC-funded research study, which assessed the outcomes of the project through the use of the sustainable livelihoods framework by comparing the status of households' assets before and after the project. The research sought to understand the diverse set of livelihood strategies pursued and their impacts on communities' coping capacities.

The methods adopted for collecting the data included a literature review, questionnaire, group discussions with some community members, meetings and interviews with some stakeholders from the government and NGO sector, and observations. The meeting with the farmers' committee focused on drawing a comparison of community conditions before and after the project, identifying the areas of outstanding project success and the underlying factors behind that success, and the current problems and risks. This section is based on the responses to interviews and meetings; all of the respondents were familiar with, and most of them benefited from, the KARP project.

Overall, the research established that, after the project, agriculture and herding remained the predominant occupations. Engagement in agriculture had increased by about a quarter above the level before the project began, with an eightfold increase in the proportion of those engaged in manual labour. Trading appeared as an occupation after the project intervention. Forty per cent of the heads of households surveyed were involved in more than one occupation, a fourfold increase from before the project. While secondary occupations can be seen largely as a coping mechanism for supplementing household income, the availability of increased opportunities for other economic activities confirms the success of the project interventions in increasing production (and hence the need for additional waged labour) and access to credit (enabling people to set up in a trade). Obtaining direct access to the vegetable market in Port Sudan resulted in increased production and an improved competitive position. The majority of people reported increased incomes as a result of the project, leading to many more people putting themselves in a middle income group. The two main factors that contributed to this improvement in financial capital were the increase in agricultural production and the new skills acquired by farmers (including soil and water management) which helped them access new resources or better utilize their existing resources.

Respecting local priorities through project flexibility

A key finding in the views expressed by the respondents was that the most important benefits to the area from the project were the two adult literacy centres and the two basic (primary) schools at Hanoweit and Ribaiet, constructed in collaboration with government authorities and a private donor. Neither of these initiatives was included in the original document for the project, which was primarily focused on agriculture and food security. The ability of the project to offer support for local initiatives not initially within the project design enabled spontaneous activities that have contributed to more secure livelihoods and greater adaptive capacity. The contribution to both education and adult literacy were positive project responses to community demands. The flexibility of the project to respond to the community's expressed wishes contributed to the positive engagement of the community with the project administration, and has resulted in positive outcomes in terms of a much more literate population.

Other enabling factors that contributed to the project's success were the cohesive nature of the community. The area was occupied by one tribal group with a culture rooted in the locality; this reduced conflicts of interest. Additionally, the tribe adheres to customary laws and traditions (*salif*) that very clearly define social rights and responsibilities and the mechanisms for conflict resolution. The fact that all the land was demarcated and registered meant there were no land tenure problems. This is an unusual situation, since in many development and adaptation projects, lack of access to adequate land resources is a key contributor to the vulnerability of the poorest households. In comparison with other parts of the Red Sea State, Arba'at community is relatively more stable and has a culture of self-reliance, cooperation and collective work. The community thus had some strong attributes contributing to their adaptive capacity: access to productive land, strong social networks within the tribal context and a culture of cooperation.

Technologies

The post-project research showed that skills and knowledge acquired during the project had been sustained. The success of the introduction of the date palm in home gardens and new improved vegetable seeds is indicated by the spread of home gardens without project support. Confidence that farming could be productive in the area also led to the digging of wells by people for irrigation in areas not served by the project. The improvement of agricultural production helped the local population stay on the land and reduced out-migration almost to zero over the project period. At the same time, building skills in accessing the market and formal credit institutions built confidence and was transferable, enabling people to develop new economic activities of their own. In this way, adaptive capacity was increased through diversification and through extension of social networks to financial and market institutions. For example, several local shops developed as a result of project learning, but independent of additional financial support.

Training

In traditional communities where the level of education and exposure to outside influences are low it is important to proceed gradually when introducing new ideas, an approach which SOS Sahel adopted. Using such an approach allowed the project administration to gain community trust and readiness to engage in new activities and practices such as women's programmes, market-orientated production, and engaging with formal credit institutions. According to the baseline survey very few community members had ever received any extension before SOS Sahel's arrival in the area. The project's training and advice sessions reached nearly two-thirds of the people, who commented positively on their usefulness.

Social networks

Developing community organizations was a vital contributor to the project's success. In particular, the committee for the distribution of water among the nine sub-villages was significant. Although community meetings used to be organized before the project, they were primarily of a political nature, initiated by the state (e.g. Popular Salvation Committees) and were irregular, dealing with issues that are of little direct concern to the community's needs. During the baseline survey, a large majority of people stated that there were no community meetings. The formation of local and central committees through a democratic process that secured the representation of all areas and social groups in the area enabled the committee to lead the community, because the community had been involved in both the planning and the implementation process from the start. Committee members thus had the confidence of the community in performing their tasks. This underlines the importance of a thorough process of organization building. After the establishment of the new committees, three-quarters of respondents attended community meetings, one-third on a regular basis; most of these people are members of one of the permanent committees. At the time of the case study research, all ongoing activities (in agriculture, adult literacy and revolving credit) are run by the elected Arba't Development Committee, which coordinates advice and support with government departments.

The formation of local organizations was very important for sustainability. According to three-quarters of the respondents, the farmers' committee created by the project has been effective and now the farmers' union is one of the four local unions forming the State Farmers' Union, despite its small size compared with the other three. It is seen as an official grass-roots authority, delegated by the community to communicate their messages to decision makers, and with the ability to lobby against the heightening of the Arba'at dam.

An unexpected outcome of the project was the recognition by local men of women's roles and rights. This was the result of bringing men and women together for the first time and allocating resources for women which led to their improved skills and knowledge, which in turn increased their contribution to household income. Women's involvement in committees represented a real breakthrough, taking into consideration the prevailing negative perceptions towards women's engagement in public life in that part of Sudan. This enhanced role for women will undoubtedly have very positive impacts on household well-being and is vital for improving a community's adaptive capacity, since women have the primary responsibility for household food security. Yet it is significant that the development of a separate women's programme was undertaken only as a result of a suggestion following the first project evaluation.

Partnerships are key to sustainability: the partnership built from the start between the local community, the project and government institutions (Ministry of Agriculture and the Agricultural Bank) ensured government support to the project when SOS Sahel funding stopped. The engagement of the state government from the start has proved very valuable in ensuring continuity of the support initiated through the project, though many local people considered that SOS Sahel had handed the project over rather prematurely. The issue of sustainability should also be considered beyond the project locality (the project site in Arba'at) to include the new options and opportunities available outside the immediate vicinity that ultimately bring in resources to support sustainable livelihoods within the area.

Weaknesses

In terms of building adaptive capacity in a region prone to repeated droughts that may well become more severe in the future, the review of the project concluded that natural population increase, in tandem with the limited land area, will certainly one day create a food gap even with good production and in the absence of severe impacts from climate change. The fish resources in the Red Sea are underexploited and their potential to provide food and an income could have been investigated. On the other hand, the Beja tribe live according to strong traditional culture, in which fishing and the sea have as yet played no part. It might not have been easy to engage them in such a new livelihood venture, with all the challenges of new skills, equipment, preparation and marketing of the fish. Awareness raising with the community about the possible future limitations to their current livelihood patterns would be a first step in opening up the possibility of fishing as an option: there is potential for development of the sector in a way that could contribute to improving the livelihood of local people.

The encouragement of the use of commercial chemical pesticides and insecticides raises questions of sustainability. On small plots growing vegetables intensively, people were keen to maximize production. However if the project was running now, encouraging the use of fossil fuel intensive inputs in a time of rapidly rising oil price would be seen as maladaptive. In general, it is important to develop the skills and knowledge for independence from potentially costly inputs, and training in the use of organic pesticides and insecticides would have been preferable.

Though SOS Sahel's success in the project was due to its comprehensive approach to natural resources and the expressed needs of the community, its perspective was largely agriculture focused and geographically limited to the Arba'at Al Zira'a area, with its very particular resources and suitability for small-scale agriculture. This limits the project's scope for replication. It will always be a dilemma for adaptation projects, which must focus on the local context to be effective, how to provide solutions that work locally while also providing lessons in process and technology transfer that are applicable more widely. With the enormous challenge that climate change adaptation poses, project design must somehow bear these twin objectives in mind if adaptation programmes are to go to scale.

A major concern expressed by many people in Khor Arba'at was the impact on their farming of a planned state government project aiming at heightening Khor Arba'at Dam to divert more water for urban use in Port Sudan. There was the risk of the weed tree species *mesquite* spreading onto fertile land and the consequent displacement of families – undermining most of the project's achievements. At the time of the research study, the farmers' union intended to lobby government on the dam and had concrete proposals to reduce the potential threat posed by the dam heightening. This involved the government guaranteeing the Arba'at population a regular share in Arba'at waters or providing support for the exploitation of groundwater by digging wells and installing pumps. The research study concluded with the hope that, since the dam project could only meet the needs of the town's rapidly growing population for a few years, the government would look to a longer-term solution.

Conclusion

The introduction of agricultural extension and improved agricultural practices in water harvesting and food crop production were effective in improving livelihood conditions in Khor Arba'at area and enhancing the overall resilience of people in the face of harsh climatic conditions (drought). Literacy education for women has led to their increased economic activity. Diversification of income through sales of vegetables and value-added crops has buffered many families from climate variability. Moreover, the accessibility of markets has shifted the weight of production for local consumption to production for marketing in nearby towns such as Port Sudan, and this has substantially reduced the out-migration and encouraged people to remain on their land. Their improved capacity to cope with the existing harsh conditions and frequent climatic variations may in the long run contribute to adaptation.

The control of the Khor waters and the registration and management of land have been the key factors that underlie and shape the resilience pattern in the area, helped by the homogeneity and prevailing spirit of cooperation among the community. These were further enhanced by the organization and capacity building of SOS Sahel's intervention. The flexibility of the project to respond to the community's expressed wishes effectively contributed to the positive engagement of the community with the project administration, and as a result the education-focused activities have clearly resulted in positive outcomes in terms of a much more literate population. Education is key to building human capacity and as such underpins many elements of adaptive capacity.

The project successfully expanded the communities' social networks, by supporting improved community organization, in particular the Arba'at Development Committee, and by building links with government departments that have continued to support the project activities with extension and improved information. Indeed, the involvement of government from the outset has been a significant factor in ensuring sustainable links with the community, and this was a unique arrangement not only for the Red Sea State but for most parts of Sudan. Through enabling access to credit, and linking people directly with the Bank of Sudan, people have been able to expand income generating activities beyond those introduced by the project. Direct and frequent contact with urban markets and their membership in the State Farmers' Union has also markedly extended the Beja farmers' social network. Access to climate information through links to the meteorological department will, if maintained, help inform decisions on livelihood options, particularly if reliable seasonal forecasts are available in the future.

Although there were some weaknesses in the project design, there are transferable lessons from the project, including:

- strong links with the state government, established at the outset;
- the lobbying role adopted by the local farmers' union;
- responsiveness to local needs beyond clear livelihood strengthening interventions;
- building skills for market access and accessing formal credit institutions, thus building confidence and social networks, as well as access to external resources.

Overall, the future livelihoods of this community under climate change depend on available water resources. On the one hand, this depends on the nature and extent of climate change, while on the other, more immediately, it depends on a government decision on the Khor Arba'at Dam. The government's proposal indicates the reality facing many marginalized rural communities: that policy favours the urban and the better off, at the expense of the rural poor. It highlights a real concern for adaptation programmes: no amount of capacity building and effective natural resource management can support adaptation in the face of policies, whether at local, national or international level, that undermine rural food security. For long-term adaptation, a supportive national policy environment and participatory decision-making framework, where communities can help shape development plans, is essential to ensure a balance between the needs of different communities, urban and rural.

Note

The research for the case study was undertaken principally by Dr Hassan Abdel Atti and Hashim Mohamed Elhassan. Dr Balgis Osman-Elasha was the principal editor and reviewer of the case study report. Further details of the research can be found in Abdel Atti and Elhassan (2003).

CHAPTER 9
Extreme weather in the Peruvian high Andes

Prepared with Juan Torres, Climate Change Adviser, Soluciones Prácticas, Perú

Abstract

Communities living in the high Andes have long experience of coping in a harsh environment, yet climate change is bringing new challenges that threaten longstanding livelihoods. Increasingly erratic rainfall and severe weather including extreme cold spells, droughts, high winds and fog are undermining potato cultivation and alpaca rearing, despite the region being the origin of and an internationally important centre of diversity for both potatoes and alpacas. This project sought to build on a successful community extension service introduced to the area by Soluciones Prácticas that creates a knowledge sharing and training network between isolated communities. This enabled the development of participatory research into pasture improvement, animal health, water conservation, potato yield and conservation of agricultural biodiversity. By supporting formal, community-based institutions, the project enabled interaction with stakeholders outside of the locality, extending the communities' ability to access knowledge and resources.

Introduction

Peru is a country with a wide range of ecological zones, including a desert coastal strip, the Andes mountain range, and tropical rainforest in the interior. Most of Peru's population lives in a narrow belt of semi-desert along the Pacific coast, supplied by water from rivers carrying glacier meltwater. Glaciers in the Andes are melting at a rapid rate and within 20 years severe water shortage in the cities is anticipated. The high Andes, whose geography consists of high plateaus, valleys and mountains over 3,500 metres above sea level (masl), have experienced significant weather changes. Always a region facing a challenging climate, rainfall is becoming more erratic, and severe weather, including cold spells with frequent frosts, hailstorms and fog, is being experienced more often. The whole country is strongly influenced by the El Niño Southern Oscillation phenomenon, which generates a marked change in rainfall patterns in El Niño years.

The Andean region is one of the areas of the world in which agriculture began, and is the origin of a number of crops of huge importance for human food security – potato, tomato, cucumbers, kidney beans and varieties of maize – as well as of several relatives of the camel: alpacas, llamas and vicunas. The province of Cusco in the southern Andes is a centre of diversity for potatoes and alpacas. The communities in these high Andean regions live with climatic extremes: intensely hot days and night-time temperatures that drop below freezing, creating high and continuous risk for agricultural production. The main assets of community members are extensive areas of land covered by a variety of grass species, providing a combination of good and coarser pasture for alpacas, and a large number of scattered natural surface water resources.

The project described in this chapter was implemented over a two-year period between March 2006 and March 2008 in Canchis, Cusco. The project area was in the valley of the Salcca River, a tributary of the Vilcanota River, falling within the districts of Sicuani, Marangani and Checacupe, at an altitude above 3,800 masl. This area is known as *puna* (high altitude rangelands) and is characterized by an exposed and bleak landscape, whose natural vegetation is coarse grassland with few trees, with many small water bodies. The project was one of seven implemented as part of a block grant from the European Community; all projects focused on some aspects of adaptation to the challenges posed by climate change in different regions of Peru.

The work presented here involved a community dedicated to cultivating native potatoes and rearing alpacas. Families depend almost entirely on their alpacas to live. They provide milk and cheese that is full of essential nutrients. Their fibre, which is extremely insulating, is used for clothing and bedding, and sold to pay for food, clothes and schooling. Manure provides fuel to help people keep warm and for cooking. Without alpacas, villagers have no means of transporting their goods (alpaca fibre and potatoes) for miles across mountainous terrain to the nearest market. Nor can they bring vital medicine and food back to the village.

The harshness of the climate means that the growing season is quite short, and only a limited range of crop species and livestock breeds can be farmed. The key problem of recent climate change for the region is the upsurge in extreme weather events (snowstorms, frosts, hailstorms, droughts, strong winds and cold spells) which affect farming activities, especially rearing of alpacas, increasing the incidence of sickness and mortality, as well as affecting natural resources essential for the alpacas such as pasture and water resources. The increased frequency of severely cold weather is significantly affecting livelihoods. Alpacas are poor foragers and struggle to find food in the snow and ice. In the extreme conditions, pregnant alpacas miscarry, meaning that the recovery of herds can take up to two or three years.

For cultivation of the potatoes the main negative impact is an increase in temperatures, an issue with potential regional or even global impact, as this particular region is one of the sites of greatest genetic diversity for native species of potato. The species needs a temperate climate as growth and harvest times are principally dependent on temperature. Above 17°C fewer tubers are formed, and below 0°C there can be severe damage to the crop. While global warming may lead to higher productivity in regions where low temperatures currently limit production, growth is currently optimal in Peru meaning that harvests are likely to decline over the entire 170 million ha of rain-fed potato cultivation.

In recent winters highland communities have faced the *Friaje*, a phenomenon of intense cold never previously experienced, with temperatures as low as -35°C. Fifty children died, and as many as 13,000 people in the region suffered severe hypothermia, bronchitis and pneumonia; 50 to 70 per cent of alpacas perished and many more were left exhausted and prone to disease. Potato crops were ravaged.

Community-based adaptation in Andean Peru

For more than 20 years Practical Action has been working in the Cusco region with farming communities living at between 2,500 and 3,300 masl. In the winter of 2005 to 2006 an extremely long cold spell brought the need for emergency support to communities living higher up the mountains, and Practical Action developed some emergency interventions focused on fodder and cold weather shelters for the alpacas belonging to the communities. This experience led to a decision to obtain funding for further work to understand how these high altitude communities, already living with climatic extremes, were coping with increased climate instability and to identify adaptation strategies. The project area was selected because it is considered to be a centre of origin for native potatoes and livestock such as alpacas, whose diversity has been developed by and is part of the traditions of the indigenous people of this region (the Aymara and Quechua).

The extreme vulnerability of the communities to climate is summarized by the impacts on the alpaca, outlined in Table 9.1.

Assessing climate knowledge and awareness

A baseline survey was conducted largely through interviews with groups, including women and men, older people and youths. There was a lack of independently sourced quantitative meteorological information for the area, and what existed was from secondary sources. However, since local people were very knowledgeable about their environment, their information was regarded as reliable. The survey covered changes in weather patterns, impacts of the changes, and mapping of natural resources including water.

The findings were that there are different degrees to which natural resources are affected by weather patterns, the most vulnerable being natural grassland. There are pastures, dominated by the species *Stipa* and *Festuca*, which are quite

Table 9.1 Seasonality of extreme weather and its impact on livestock

Climate event	Period	Impacts	Comments
Hail	January to March	Enterotoxaemia, directly affects newborn animals, increasing mortality if not treated	Generally comes with flashes of lightning that have led to deaths in livestock and humans
Frosts	June to August; sporadically from January to March	Pastures dry out and this vital resource cannot be relied on for feeding the alpacas	
	September to October	Has a direct influence on the regeneration and/or new shooting of grasses, and in the incidence of abortions	
Snow	January to March	Illnesses caused by snow include pneumonia, conjunctivitis, queratitis (an ulcer on the cornea, caused by bacteria or virus), the last of which is seen in people as well as in alpacas	If not excessive, snow has a positive effect on the quality of the wool
	June to August	Combines with rains and frosts	
	September to October	Plays a key role in regeneration and new shoots of grasses	
Strong winds	July to September	Cause abortions, representing the fourth most significant risk factor for rearing alpacas after hailstorms with frost, snow with frost, and frost with drought	
Heavy rain	Starts December, through to March	Lower risk. The permanent humidity generates conditions for sickness (diarrhoea) developing resistance to treatment and elimination of the illnesses	
Droughts	August–October Generally January to March	Critical period for sufficiency of pasture for livestock. Shortage of grass leads to a process of acute debilitation, which affects particularly female alpacas and their young. During drought thin and sickly animals die	

coarse and cope better with frosts and droughts. Other pastures are badly affected and further suffer from the increase in fires which follow drought and subsequent overgrazing. The level of mortality in alpacas due to weather extremes is about 30 per cent for young animals, 10 per cent for adults, while the miscarriage rate can be as high as 20 per cent. The level of loss of native potato crops due to frosts and hailstorms fluctuates between 50 and 60 per cent. In the face of extreme conditions, this figure can reach 90 to 100 per cent.

As part of the baseline survey for the project, testimonies were taken from men and women over 40 years old who have lived their entire lives in the high Andean lands; they were able to compare what they were seeing now with their memories of when they were children more than 30 years ago. The testimony from one couple was particularly detailed and vivid and is presented in Box 9.1.

In summary, the most important changes being experienced are as follows:

- *Reduction in snow cover and the dwindling of glaciers.* The Vilcanota peak and the surrounding ranges, which 30 years ago were almost permanently covered in snow, are increasingly remaining bare of snow.
- *Reduction in water sources.* At higher altitudes the principal sources are springs, lakes and brooks. The springs have begun to reduce in flow by up to 40 per cent, and in some cases have dried up completely.
- *Rainfall pattern.* Rains were previously more or less uniform (December to March), but now feature abrupt variations: there are strong and plentiful rains at unpredictable times and drought at others. Hail and thunderstorms are more extreme than previously experienced.
- *Intensity of sunlight.* Sunlight has become more varied; there are some hours of abnormal heat, but mornings and evenings have become colder. One respondent noted that 'the cold is freezing, the heat burns you, clothes dry so fast they smell of burning'.
- *Winds.* Wind is now strongly concentrated between the hours of 2 and 5 in the afternoon: 'the wind is so strong, you can't work, we can't carry on our normal shepherding work'.

A significant aspect is that extreme weather events are coming at quite unexpected times of the year. It is these out of season events that seriously affect crops and livestock and have led to an increase in acute respiratory problems in children and old people. The challenges of climate change are overlain on degradation caused by human activities, including overgrazing, burning of pasture and drainage of ponds; the changes which these activities bring to the environment themselves initiate processes of desertification which create conditions for local micro-climatic changes, different from those caused by natural climate variability, and particularly so over the last 30 years.

There is a strong tradition of living with climate variability in these communities, evidence of which is the large number of natural signs used for predicting weather over short periods (examples of which are described in Box 9.2). These indicators are either biological in character (relating to behaviour of plants, birds and insects), or physical (relating to the moon or stars). This local knowledge is used for planning, breeding and management of crops, water and pests as in the mountains there are no meteorological stations or other scientific means to predict climate. The people learn from their early years to adolescence to observe natural signals and to correlate

> **Box 9.1 Experiences of Andean Alpaca herders**
>
> Juan Faustino Yucra, 52 years old, from Chillihua and Rosa Isabel Supho, 37 years old from Tañihua, report that the natural environment has been gradually changing:
>
> > Nowadays the glaciers on the mountain tops have disappeared; precipitation has turned into hail accompanied by thunder and lightning and not snow as in previous years. Cold spells extend up to September, when previously it was normal for this phenomenon to occur only in June and July. Winds are more intense and temperatures are higher during the day with extreme heat, and they descend lower than before at night. All these changes have brought increased illness to people and livestock.
> >
> > The absence of rains, high temperatures and low availability of water at this time of year [September] has caused the pasture to dry up too much and become tinder dry. The winds blow loose hay away, leaving grazing areas with little dry vegetation.
>
> The delay in rains brings the late flourishing of pasture and, put together with the presence of freezing conditions in January and March, threatens availability of pasture for animal feeding.
>
> In response to these climatic difficulties, which Juan and Rosa think are irreversible, they are protecting themselves from wind and cold; they are working on maintenance of irrigation canals to make more efficient use of water, using water in 'doses' for each plant; and they are working to improve their knowledge of mechanized irrigation techniques.
>
> Another alternative that they see as a partial solution is to reduce their herd sizes, as overgrazing also contributes greatly to desertification and the challenge of feeding the animals becomes ever more demanding.
>
> This couple voiced the anxiety of all the alpaca herders: this is the only life they know, yet they can only carry on doing it as long as nature allows.

these factors to make a judgement on whether the planting time, season or year to come will be favourable, regular or unfavourable for agriculture or livestock. Most of the signals relate to precipitation: whether the rains will be early, on time or late and whether there will be unusual variations like hail, snowstorms or drought. Alongside the traditional signs, people now have access to information about local weather from radio. The local farming and weather bulletins and those from the Ministry of Agriculture at the national level emphasize the relationship between climate and annual or perennial crops. Local radio also broadcasts information on indigenous signs and local people use the biological indicators rather than meteorological information for taking their farming decisions.

Radio is important for communicating climate risk in the mountain-top zones. People can get weather forecasts, information on the El Niño effect, or about a cold wave closing in. Because of such forecasts in 2008, for example, an emergency situation was declared for the mountain-top communities in Cusco.

> **Box 9.2 Traditional forecasting knowledge in the Andes**
>
> *Reading the clouds*
> The clouds and rain in the first days of the month of August are carefully observed. One method is to look at only the first three days: every day stands for one month (1 August is January, 2 August is February and 3 August is March). Another method (mainly used in the valley communities) uses 1 August to describe August, 2 August for September and so on up to 8 August representing March. If during the day there are no clouds that means there will be no rains in the corresponding month. If there is presence of a greater or lesser amount of cloud, either in the morning or afternoon, that means there will be a greater or lesser amount of rain in the corresponding month.
>
> *Flowering of plants*
> There are various plants which flower between August and October. The quality and size of the flowers indicate the coming climate in terms of the rains in the following season. If the first flowers are good, the rains will come early (meaning it is possible to sow crops early in the season). If the first flowers are weak but the later ones are very good, it is possible that around the first sowing period there will be freezing conditions or drought, in which case it is better to delay sowing.
>
> *Behaviour of birds*
> Birds that prepare their nests underground are watched for the way in which the soil is removed and how they place it. This indicates how successful crops will be in the coming season, and accordingly people make provision by saving foodstuffs, mainly potatoes, for the survival of the family.
>
> *The fox*
> The date the call of the fox is first heard (usually in September or October), its shrillness and the time of day are all very important in determining if rains are going to be good, average or bad.

Awareness raising

Activities to raise awareness included workshops, exchange visits, discussions and conferences. Meetings were very interactive, since the local people understand their climate and the project took place in a region where agriculture has been practised for millennia. Presentations included slide shows, pictures and diagrams as well as annotated maps. The content focused on climate variability in the high provinces of Cusco and the signs that have been appearing for 30 years indicating that the climate is changing, such as the reduction in the snowline around the high provinces of Cusco and the new diseases appearing in livestock.

Through awareness raising, the project also built the capacity of the community to submit proposals in accordance with the planning processes of local government in the districts of Sicuani, Maranganí and Checacupe. The community sought support in adapting to climate change, and were able to have their needs included in local government budgets. Crucially, budgets are prepared with the participation of communities as part of the decentralized local planning process in Peru.

Social networks

Building the capacities of community members was considered the most important target for guaranteeing the sustainability of the project work, and for this reason all activities included training as a central element. Farmer field schools were a key method and focused on building leadership capacity by using previously trained farmers to train new farmers. These people were able to reach out to the farmers in their own language and with their own terminology. This training of 'Kamayoqs' has been developed by Soluciones Prácticas (Practical Action's office in Peru) over some 22 years and has led to the emergence of a group of indigenous technology leaders using a training approach that complements the cultural and social context of local farmers. The training enables a two-way flow of information between those promoting development and the local farmers.

The Kamayoq have become the promoters of appropriate technologies, generating creative solutions to local problems. The technologies and practices that the Kamayoq learn and subsequently teach to others are not always new, but access to this information is not widely available to smallholder farmers. In 1998 the trained Kamayoq had formed a legally recognized *Asociación de Kamayoq de Canchis*. The Association has a building in Sicuani and a technology library that was established under an agreement with the provincial municipality of Canchis. By 2008, six rounds of training of Kamayoq had taken place.

Training is conducted increasingly in the local language, Quechua, in order not to disadvantage local people, and especially women, by teaching in Spanish. Women have particularly benefited from being trained as Kamayoq and have focused on activities that are important for them: raising of guinea pigs and chickens, animal health and vegetable production.

Evaluations have shown that the impacts from the work undertaken by the Kamayoq include:

- increased income from increased quality and quantity of food production;
- improved well-being (confidence, diet);
- reduced vulnerability;
- improved food security;
- more sustainable use of natural resources.

The education focus of the work is based on the principles of the educator Paulo Freire, and is characterized by being: *decentralized* – training takes place in a variety of locations such as farmers' plots and classrooms; *practical* – the content of the training programme complements farmers' needs; and *culturally sensitive* – the training is predominantly in the local language, Quechua, and is respectful of local customs and norms.

The development of technologies in this project drew on the work of the previous 22 years with local communities and Kamayoq, and the capacity building for adaptation followed the same model.

As part of a training programme on potato cultivation two exchange visits were organized where the newly trained potato experts shared their knowledge with more than 600 potato growers from 20 communities in Canchis province. Networks were also built through the graduation ceremony for the newly trained Kamayoqs, where senior officials from the Ministry of Agriculture, officials from local municipalities, and the Association of Kamayoqs were all present, ensuring the new graduates were able to develop links beyond their locality.

A key strength of the mountain communities is a long tradition of community organization of peasant farmers, in units called *ayllus,* based around small settlements. *Ayllus* are a pre-Columbian form of social organization, comprising groups of families who have a degree of kinship and live in the same locality. A system of reciprocal working exists between members of the *ayllu*, called *ayni*. *Ayni* is mostly focused on agricultural work and house-building. A group helps one family on condition that they will reciprocate when requested. *Ayllus* work on communal lands as a group, although people have individual parcels of land and own their own houses. There were also other civil society organizations operating in the project area: farmers' groups, local government and local cooperative associations.

One of the central elements was not only training existing organizations about climate change, but also creating new organizations called 'Community committees for risk management'. These are community-based organizations which aim to confront threats that can arise as a result of drastic climate change. Their role is to work on prevention (by informing and preparing people for possible events) and recovery from extreme events. The project ensured that women and men were equally represented on the committees. Moreover, the formation of 22 community civil defence committees among the mountain communities was facilitated by the project. These committees are integrated into the national civil defence system, so that in the case of emergencies they coordinate with regional governments, local governments and communities, to mobilize aid or other actions. At present, however, there are still no disaster risk prevention information systems specifically designed for the high Andean communities.

The project results were presented to decision makers in municipalities, mayors, departments within the Ministry of Agriculture (National Programme for Managing Water Catchments and Soil), academic bodies, and related organizations such as the National Meteorological and Hydrological Service and the National Civil Defence Institute. The result has been a much increased interest in and understanding of the impact of frosts on agricultural productivity, and the role of the new local civil defence committees in preparing for severe frosts.

Technology development

Local knowledge and modern technologies together shaped the activities developed for adaptation and reduction of vulnerability to climate change. For the livestock keepers, the project had three components linked to different elements of the ecosystem: grassland improvement, health of alpacas and water resources. Fifty women came forward for training, of whom 44 were selected, and 25 completed all sessions to become Kamoyoqs, specializing in alpaca rearing – including breeding, animal health care, management of pastures and converting alpaca wool into craft products. Training focused on livelihood strengthening rather than the usual Kamayoq model of providing a technical service to the wider community. A further 25 people were trained as experts in native potato cultivation, in a process of nine modules of two or three days over a period of several months. The training encompassed field cultivation, handling of seed potatoes, post-harvest management, processing and commercialization. In addition to the training, various research trials were undertaken, as described below.

Pasture improvement

Three technologies for pasture management were implemented during the project, all with the aims of improving pasture to increase the productivity of the alpacas and building resilience to an increasingly harsh climate. The first was the establishment of demonstration pasture plots of different seed mixes and compost additives; the second involved using techniques for minimum tillage to improve high altitude pastures; and the third introduced techniques for the careful use of water resources for the irrigation of grassland.

Reseeding of degraded pastures is a key tool in responding to climate-affected or overgrazed pastures. To assess whether different species can adapt and be productive at altitudes greater than 4,000 masl, trial blocks were established, within which a single species was sown on six plots and mixes on eight plots. Trials included new species such as English rye grass and Italian rye grass, while other trials used grass mixed with alfalfa (*Medicago sativa*) and clover (*Trifolium perenne*). Using this method species mixes were selected for three different ecological zones.

Another series of experiments analysed the most effective way to improve soil fertility and evaluated the impact on potato yields, using trial plots with control plots next to each. The conclusions were that mixes of grass species do best, with productivity 25 per cent higher than monocultures; productivity was further increased when organic compost was added. No tractors or other mechanical cultivation tools were used – just traditional hand tools and animal-drawn ploughs.

Conservation of natural grasslands, to make them sustainable sources of food for livestock in conditions of climate change, requires continuous management by the alpaca herders of the high Andean zone. Although the plants can in theory cope with the pressure of constant grazing, the carrying capacity is currently being exceeded, and this, coupled with extreme weather affecting the growth and productivity of the pasture, requires the development of technologies new to these communities. In response, 10 areas, each of 50 hectares, were protected by erecting wire fencing to prevent grazing and allow pasture recovery. Within these areas, a variety of grasses and grazing plants were incorporated by seeding into the natural pasture, and their growth encouraged through use of sprinkler irrigation. The new growth and reseeded natural forage plants in the enclosures ensured seed production, and these new seeds were subsequently dispersed more widely in the wind and through animal dung (since the seeds are not digested by the animals). The protected parcels of land with good vegetation cover also helped retain moisture in the soil, preventing the natural pastures from becoming eroded.

Development of animal health technologies

Rearing of alpacas in Canchis province has been badly affected by *Fasciola hepatica* (liver fluke, causing the disease fascioliasis). The disease formerly only affected cattle and sheep at lower altitudes, but is now present at high altitudes through a combination of factors: cattle are brought up from the valley with no health controls and are allowed to mix with alpacas, and the warmer temperatures now being experienced at high altitudes enable the parasite to survive. Communities at higher altitudes also suffer because of their remoteness, poor access to veterinary services and their extreme poverty.

The traditional strategy to protect animals against health risks and control mortality was the use of herbal medicines, but the erosion or devaluing of traditional knowledge has led to the use of commercial chemical preparations. Even for the 10 per cent who can afford the cost, the use of chemicals generates dependence on external inputs which can increase vulnerability. Practical Action proposed laboratory testing of the natural medications to determine whether those used on cattle would be effective in alpacas. The ingredients consisted largely of the root, leaves, petals and seeds of several locally growing plant species. Similar tests were also carried out on treatments for intestinal parasites. This process validates local knowledge and enables the traditional medications to be more widely shared. The results showed that, with an appropriate dosage, the natural medicine was reasonably effective: 70 per cent of animals in the experiment were successfully treated with natural medicines, compared with 100 per cent for the chemical treatment. However, the lower efficacy of the natural medicines is offset by their affordability. For maximum disease control, medication must be combined with the rotational use of pasture so that animals are not continually reinfected.

Water conservation technologies

The natural water resources comprise streams and wet hollows (*bofedales*). The high number of animals using the pastures, coupled with a drying climate, has led to a gradual reduction in available water. Two strategies were developed for conserving water: first, the conservation of wetlands through enclosures, and second, the introduction of improved sprinkler and gravity irrigation systems which drew water from streams via small-scale piped channels. The works were carried out in 10 communities, benefiting 350 families. Nominated and trained 'irrigators' were organized into multi-family groups (irrigation committees) in four communities comprising a total of 260 user families. Capacity building was provided for 318 families of livestock keepers on managing and maintaining small sprinkler systems for improved pasture.

Improving potato harvests

Three strategies for adaptation were tested for the farmers of native potatoes: 1) research factors that could improve harvests suffering under the changing climate; 2) conserve the diversity of native potatoes; and 3) raise awareness about the pests and diseases that are likely in a changed climate.

Native potatoes are the only varieties able to grow at high altitudes. Participatory research was carried out to examine different ways of increasing yields in the face of more frequent frosts that are leading to smaller tubers. Experiments were designed to investigate how to increase the harvest by planting different size tubers at varying densities. The experiments revealed that the highest yield per hectare was from medium-sized tubers, increasing yields by almost 30 per cent (5,880 kg per ha) compared with the larger tubers. This is because medium-sized tubers can be more densely planted. With reduced rainfall, smaller tubers were found to produce the best yields.

Another area of research was the use of compost and manure collected from the night-time enclosures. The experiments were developed by project staff with a group of 14 producers, 11 men and 3 women, by cultivating potatoes using different types of soil improver on land that had been fallow for six years. During cultivation, weather events, pest infestation, size of plant and other variables were monitored. Pest management was carried out by manually removing weevils. The best results were obtained from well-rotted compost and the worst from ash (a traditional source of potash). However, all harvests were significantly greater than those where compost was not used. The outcome was that people were encouraged to make compost, well in advance of needing it, so that it would be well rotted. Various other organic agricultural practices were introduced, including biological control using beneficial insects and natural plant-based pesticides and insecticides.

Conservation of agricultural biodiversity (native potatoes)

The Andean communities have been guarding their potato diversity for centuries. Key seed keepers in a community can be cultivating 900 varieties on their land. As a result there was little that the project team could do to supplement the traditional methods, which were observed and recorded. Certain parcels of land have very great biodiversity. This is conserved by sowing different varieties each year in slightly different terrains and altitudes, so that if the crop is lost in one place because of, for example, a severe frost, the crop might survive well in another place. Wild relatives of cultivated plants are conserved where these are found near the cultivated parcels. There is a strong tradition of conservation of seed varieties through exchange of potato seeds at harvest time. Seed is exchanged with different communities even as far away as Puno (approximately 390 km away), creating Cusco–Puno biodiversity conservation corridors. This practice renews the gene pool. Many traditions and rituals surround the potato harvest, including offerings to Mother Earth (*Pachamamma*) which promote respect for nature.

Resource implications

The project team comprised eight people, for a beneficiary population of 1,326 direct beneficiaries and 3,978 indirect beneficiaries, scattered over an area of 83,603 ha. The total project cost was €103,500 ($151,200; exchange rate €1 = $1.46, 18 December 2008), implemented over 2 years, around €80 per direct beneficiary. A direct beneficiary is a farmer who received technical training and advice from a Kamayoq. Indirect beneficiaries were the other members of the household, benefiting from increased productivity as a result of the training.

Lessons and challenges

Local knowledge

Extensive knowledge of local weather and biological indicators, the area's natural resources and how to manage them, and their skills in conserving agricultural biodiversity will be of benefit to the community as they seek to adapt to climate change. The long tradition of observing weather and correlating it with other biological events and behaviour is a skill that relates closely to those needed for scientific experimentation, illustrated by the efficacy of the traditional medicines developed in the communities living at lower altitude (but of the same Quechua tradition).

Social networks

The project built on two important characteristics of the communities. Firstly, their traditional organization into small hamlets where cooperative working was customary facilitated joint working on the project and the use of participatory research methods for testing improved technologies. Secondly, and at least as important, the already well-established and familiar system of farmer promoters, the Kamayoq, provided a way of linking the community with outside knowledge on agriculture in an affordable and accessible manner.

Although the networking within communities was quite extensive, their remoteness and the lack of transport infrastructure meant that links to outside institutions and knowledge was limited. The benefits of establishing local committees to link with the regional and national network of civil defence committees was a valuable step in building links with government for support during emergencies, which may have wider benefits in terms of relationship building with other local government institutions.

Training

For people with a low level of literacy, learning by doing is the most effective approach for acquiring new skills and knowledge. Development of demonstration plots was found to be a valuable approach for testing and promoting new technologies for crop management, improved grass mixes and alpaca rearing. Pilot demonstrations of small irrigation structures and fodder storage buildings were also constructed.

Undoubtedly the success of this short adaptation project was a result of its strong foundation of working on a community-based extension model with proven success. This meant that trust with the communities already existed, and there was a strong shared cultural understanding between project staff and the communities.

Technologies

The project activities built on the assets of these communities, in terms of potential for adaptation to climate change: a large diversity of native potato species; extensive unimproved pasture; and long traditions of management of the few locally adapted crops and livestock. The focus in terms of technologies was on strengthening these assets to overcome the decreasing productivity of both potatoes and alpacas as a result of increased climate variability.

The validation of natural medicines developed for cattle as an effective treatment for alpacas was important not only for increasing the affordability of medication for a debilitating illness, but also for increasing people's confidence in the Quechua culture's store of traditional knowledge. It also served to demonstrate the scientific basis of traditional knowledge to agricultural advisers involved in the experiments, a way of building a bridge with other institutions and generating respect for communities often regarded as uninformed.

Awareness raising

Sharing experiences and awareness raising on climate change issues was carried out through discussions which were lively because of the depth of knowledge people had. On available technologies for strengthening livelihoods, the most effective means of communication and learning were courses (42 were run) and exchange visits (three to other communities). Outreach beyond the project communities was achieved through local seed and breed fairs as well as communications material such as leaflets, which reached a wider audience including local municipalities, NGOs and neighbouring communities.

Challenges

Despite the successes, there were some lessons for improving future project design and implementation. Local knowledge should be better integrated into the design of proposed technologies, perhaps recovering other local traditional technologies that have demonstrated their success for dealing with climate changes, such as the local indicators of change. There could have been more linking and coordination with local institutions during implementation. The lack of locally collected meteorological information needs to be rectified including the establishment of simple weather stations throughout the whole watershed – not just on the high plateaus but including communities at the heads of the valleys – so that a wider picture of the changes taking place can be gathered.

Because the high Andean communities of Sicuani have a long tradition of relating to microclimatic variability, they could well represent a pilot area for research to monitor the processes of climate change in relation to rearing of alpacas and their relatives, which are economically very important to Peru and Bolivia. Another issue not handled in this project, but important for these isolated communities with their limited range of products, is that of market access. Increased income requires access to markets; currently in Peru the only market for native potatoes is mainly for export, but it is growing fast. Niche organic markets are in general growing at a rate of 10 per cent annually. Soluciones Prácticas is actively developing that market within the hotel industry in the Cusco region, raising awareness among top chefs, and introducing recipes. 2008 was the International Year of the Potato, and activities were organized to promote the wider consumption of native potatoes, by bringing community people and chefs to promotional events such as potato fairs.

Conclusion

The farmers and shepherds in the high Andean zones have lived with an extremely challenging climate for millennia. The area is one of the origins of agriculture and water management and has a history of 12,000 years of human habitation. While their remoteness and the low productivity of natural resources has meant enduring great poverty, it has at the same time given the communities a key advantage in terms of adaptive capacity for further climate change: they have considerable understanding of the variations in their local environment, and have developed strategies for managing climate variability. They have a store of traditional knowledge for managing agricultural biodiversity under climate variability, particularly of native potatoes, but also of their alpacas, and this can assist with adaptation to climate change.

In so far as they have a strong relationship with their local environment, close observation of weather patterns, and long experience of living with microclimatic variability, these communities are in some respects in a better situation for adapting to or living with a changed climate than communities used to a stable weather pattern. However, their extreme poverty, their remoteness from services and the harshness of their environment and climate mean their social networks are limited and they have few additional resources to draw on to cope with the increased shocks they are facing. A strong tradition of shared working supports people in this harsh environment, but for future climate changes, greater connectivity to wider social networks to access information and resources will be important in enabling the development of other livelihood options and links to markets. The development of civil defence committees for mobilization during emergencies, and the linking of these committees into a local, regional and national network is a vital first stage in strengthening social networks. A second important stage has been the training of Kamayoq from this remote community, including women, specifically in the skills of alpaca rearing; while this was not carried out under this project alone, it was a crucial element in developing successful technologies for adaptation.

Effective new knowledge was successfully introduced into livelihood strategies through participatory testing of new crops and technologies for raising productivity that used local resources and built on traditional knowledge. The manner of knowledge sharing, based on an existing successful model of farmer-led extension founded on traditional knowledge, meant that the information was accessible to this remote and strongly traditional culture.

For future adaptation to climate change, the building of human capital and social networks will be crucial. In this respect, the issue of ensuring market access is important since sustainable benefits from improved production require effective integration into local and regional markets in order to provide income and a spur for further productivity improvement. Ultimately, though, while further adaptation is likely to be achievable through improved technologies, there is a fear among these hardy alpaca keepers that the limits to their ability to live in their environment could one day be reached, and this cannot be dismissed.

CHAPTER 10
Conclusion: community-based adaptation in practice

Abstract

This chapter reflects on the case study chapters to draw out lessons for community-based adaptation in practice. The first section considers the approaches adopted in the case studies against the framework set out in the Introduction. In the second section, the challenge of scaling up community-based adaptation is addressed, drawing on the learning contained in the main part of the book and finding that the principal threat to building adaptive communities is institutional – and therefore political – marginalization.

The elements of adaptation

The framework set out in the Introduction identified two factors that are significant in designing adaptation activities: vulnerability to climate change hazards, understood in terms of the broad environmental and socio-political context (that is, starting-point vulnerability); and the clarity of the available knowledge about the hazard. These underpinning aspects of adaptation play an important role in determining the blend of vulnerability reduction, resilience strengthening and building of adaptive capacity that are required in a given context, yet have received very different levels of attention in the case studies examined in the preceding chapters.

In all of the cases studies, significant amounts of time and energy were invested in establishing vulnerability to climate-related hazards: in Pakistan, for example, the process of establishing baseline knowledge continued throughout the first year. While the methodologies were different in each case study, the purpose was similar: to identify the significant climate-related hazards facing the community, and to understand the human and environmental components of vulnerability to those hazards. For most, this process involved mapping existing livelihood practices onto current and (less explicitly) future climate trends. Indeed, in the majority of cases reference to future climate predictions was limited to noting the likely worsening of the current climate and weather phenomena that undermine livelihoods. The second significant factor in adaptation – the clarity of climate and forecasting information – was only referenced in the Kenyan example, and in this case was deemed too poor to be of use. A focus on vulnerability to current weather conditions

Table 10.1 Adaptation strategies employed in each location

	Strengthening resilience *Increasing the ability to absorb shocks or ride out changes*	Building adaptive capacity *Improving the ability to shape, create or respond to changes*	Vulnerability reduction *Targeting identified climate change vulnerability*
Bangladesh	Community representation on local government disaster planning committee Alternative livelihoods and food sources	Building community leadership in different age groups Linking community, local government and non-governmental organizations Training courses on climate change issues	Early warning system training and organization Food security during flooding (e.g. floating gardens and duck rearing) Flood protection technologies (e.g. elevated housing and wells)
Kenya	Diversity of seed types to address uncertainty in weather forecasts Seed bulking and saving including traditional varieties Alternative livelihood practices including trade and manufacture supported by informal credit mechanism	Training and support for experimenting with seed selection Project created a strong relationship between the community and government arid lands project, including meteorological office staff	Food security improvement through agricultural management including new varieties and seed storage Water capture for domestic and agricultural use via sand dams
Niger	Animal loan system to enable restocking Training and equipping health care workers Participation in policy processes to enable land access and community management of natural resources	Schools and adult literacy (supporting decision-making engagement with policy processes as part of a long-term project) Building community leadership and knowledge of climate change	Water security via wells Food security through construction of grain and fodder banks Land regeneration via dams to improve food security
Nepal	Improved resource management and alternative crops (food and income diversification) Goat raising including insurance mechanism Formation of forest user groups (including a female group) and community representation on district disaster mitigation network	Growth in confidence of farmers to make strategic decisions and use alternative technologies	Flood protection through watercourse management (check dams, soil moisture retention and forest management) Alternative agricultural practices to respond to rainfall changes and increase yields following loss of land to flooding

	Strengthening resilience *Increasing the ability to absorb shocks or ride out changes*	**Building adaptive capacity** *Improving the ability to shape, create or respond to changes*	**Vulnerability reduction** *Targeting identified climate change vulnerability*
Pakistan	Diversified livelihood strategies including multiple cropping, trades and alternative livelihoods for women Training of community vets	Access to government extension services and alternative technologies through the creation of resource centres	Alternative crop production to increase resistance to drought and flood Vegetable growing to reduce dependence on water-intensive crops
Peru	Improved productivity of grasslands Improved animal health Conservation of local potato varieties	Participatory research skills applied to grassland and potato productivity	Improved pasture management to reduce vulnerability to frost More widespread use of low cost natural medicines, reducing vulnerability of livestock, a key asset
Sri Lanka	Rebuilding access to diverse traditional rice varieties	Ability to experiment with and make decisions about alternative rice varieties Institutional support for access to and selection of rice varieties via paddy cultivator group	Reducing rice crop losses to salinity through the introduction of alternative varieties Reducing coastal erosion and land loss through coastal planting
Sudan	Improved incomes through increased agricultural productivity and access to markets Diversified livelihoods	Building primary schools and holding adult literacy classes Building organizational capacity Linking the community with government departments, including farmers' union partaking in state-level decision making	Improved food security through alternative agricultural practices including small-scale home gardens Improved management of increasingly unreliable local water resources

generates adaptation projects that are principally concerned with the existing climate rather than future climate change. Most of the case studies reflect this, with work first and foremost driven by the reduction of vulnerability to current hazards (including both category 1–fast onset or transient events and category 2–incremental changes), often through the adoption or adaption of alternative technologies.

Adaptation and existing vulnerabilities

Table 10.1 provides an overview of the three components of adaptation practice in each case study, illustrating the connections between activities within projects and the different approaches that can be taken to strengthen resilience, build adaptive capacity and reduce vulnerability. One notable feature of Table 10.1 is the continuity that in many cases is formed between the entries across each row. Often, the pattern reflects a spectrum of activities, from technological changes to achieve vulnerability reduction, through mixed technical and socio-political approaches to resilience, to purely socio-political interventions to build adaptive capacity. The elements can also be seen as a continuum, in which strengthening resilience, vulnerability reduction and building adaptive capacity form a coherent intervention. However, in the majority of chapters this continuity emerges because the projects take the need to reduce existing vulnerability to climate change as the starting point. Work on resilience and adaptive capacity follow from this focus. This is clearly visible in the Sri Lanka example, where each of the components of the project reinforces the other and contributes towards a clearly defined goal. Informed by the failure of hybrid rice introduced during the green revolution decades, the approach to saline encroachment onto rice paddies is one that seeks to increase decimated yields (vulnerability reduction). The method to achieve this employs variety diversity (strengthening resilience) while developing the skills and opportunities needed to productively employ new varieties (building adaptive capacity).

In many of the case studies vulnerability reduction efforts overlap with measures to strengthen resilience. This is principally due to the adoption of 'no-regrets' approaches to vulnerability reduction, where measures are introduced that bring immediate reduction in vulnerability to an existing climate hazard without being dependent on a particular climate future emerging. Frequently, this generates approaches that also increase resilience, such as through the introduction of multiple livelihood strategies or a shift in cropping practices to a diversity of varieties, as in Pakistan and Sri Lanka. These examples illustrate the clear potential for synergies between vulnerability reduction and resilience, but it is important to note that without the attention that 'no-regrets' focuses on resilience, these synergies do not necessarily emerge. Vulnerable livelihood practices may just as easily be replaced with a trade or other income-generating strategy that is best suited to the current climate, and saline intolerant crops, for example, can equally be replaced with varieties whose yields are optimized to the current conditions. However, just as the Introduction noted that increasing resilience reduces vulnerability to the broadest possible range of hazards, so too can a no-regrets approach to vulnerability reduction contribute to strengthening resilience.

When alternative livelihoods appear as resilience strengthening it can give the appearance that promoting resilience is the same as reducing vulnerability to climate hazards. Although a similar class of activities is employed (alternative livelihoods introduced) the function is different, as this is not an approach to reducing vulnerability to a specific climate change hazard. This form of resilience strengthening can emerge for two reasons, either as a direct measure to improve resilience through diversified livelihoods, or to satisfy short-term community needs. As many of the case study chapters note, interventions will normally need to address the immediate difficulties that communities are facing in their everyday lives if an interest in the problems of climate change is to be secured and adaptation made relevant to the local population. Improved resilience can result from this if alternative livelihoods are employed to meet an existing need for improved income and/or food security (where the need is not due to the presence of a climate-change hazard). Once again, synergetic strategies can be found that simultaneously meet existing needs and strengthen resilience – but, again, these synergies will not necessarily result unless interventions specifically seek to build on the principles of resilience.

Coherence between activities in adaptation projects thus relies to an extent on an understanding of the need for resilience. However, a focus on existing vulnerability can lead to projects that do not contain distinct adaptive capacity or resilience-building measures. This can be understood through the adaptation space, illustrated in the Introduction in Figure 1.1: concentrating on known vulnerabilities means that both clarity and vulnerability are perceived to be high, and thus the focus of adaptation activities is dominated by vulnerability reduction. The possibility of uncertainty is overlooked and thus adaptive capacity building and resilience strengthening measures emerge as a consequence of vulnerability reduction rather than through an independent focus on these important aspects of adaptation.

Recognizing the role of climate and forecasting information

The Kenya project can be distinguished from the majority as it seeks to build adaptive capacity as a principal project activity. Along with Sudan, it is the only case study to include explicit efforts to identify or develop sources of climate or forecasting information. Appropriately communicated climate and forecasting information is a key element of adaptive capacity as it secures the ability to plan for (rather than only respond to) changes in the weather and climate. The Kenya example sets out to ensure improved forecasting information through the communication of seasonal forecasts. From the outset the project aimed to bridge the differing expectations, experiences and world-views of the local community and meteorological office staff, working as a boundary organization linking the two stakeholders. While climate change predictions were felt by the project coordinators to be too unclear for use in current adaptation activities, shorter-term seasonal forecasting information was sufficient to support forward-looking adaptation. Placing the need for forecasting at the heart of the project provided a focus on relationship building between the community and forecasters; fostering experimentation with seed varieties subsequently emerged in response to the uncertainty in forecasting information.

The failure in the majority of cases to recognize the use of forecasting and prediction information as part of adaptive capacity means there is no explicit engagement with the uncertainty, unreliability or statistical nature of forecasts. While all the projects examined here included activities to strengthen resilience and build adaptive capacity (and thereby implicitly address the lack of clarity in future climate predictions), doing so from the perspective of existing vulnerabilities does not necessarily lead to a recognition of the need to understand or secure access to forecasts and predictions. It may also disguise an implicit assumption that the current conditions or climate predictions represent an accurate indication of the future. This approach has fundamental drawbacks. As the Pakistan example illustrates, unexpected climate variability can undermine efforts to introduce new agricultural practices (in this case frosts damaging young plants). Similarly, in Nepal unexpected flash floods washed away newly repaired irrigation channels. In Sri Lanka, while a lack of relevant national research on climate change impacts was identified as a fundamental barrier to adaptation, change in this area was not advocated within the adaptation project. It is a lesson of the experiences examined in this book that a recognition of the need for climate and forecasting information, and an analysis of how and why existing social networks fail to satisfy that need, is an important but easily overlooked component of community-based adaptation interventions. However, any emphasis on climate and forecasting information need not be at the expense of vulnerability reduction and, rather, should be seen as part of an integrated approach to adaptation. In any context there will be different needs: the urgency of current flooding problems documented in the Bangladesh and Nepal chapters, for example, rightly necessitated focus on vulnerability reduction. Here, early warning may require improved access to short-term weather forecasts, but in conjunction with infrastructure improvements and local institutions to coordinate emergency response. In Nepal, flooding was partly as a result of changes in rainfall patterns which were also reducing yields from traditional crops such as rice. While this problem requires the identification of alternative crop varieties as part of a no-regrets vulnerability reduction strategy, it also suggests a need for seasonal forecasting to ensure the viability of new varieties in a changing climate. Each hazard type thus demands a different response and has a different relationship with climate and forecasting information.

Many of the case study chapters note that it is a challenge to engage communities in abstract discussions about possible future climate change events. However, awareness raising activities were successfully employed. Although this was not the same as communicating the complexity of climate predictions, these activities provide an effective first step in bringing climate information to communities and explaining the long-term process of climate change. The case studies illustrate a breadth of awareness-raising activities; the Bangladesh and Pakistan projects in particular tailored the communication medium to target, in different activities, a population with very low literacy, an indifferent civil society and extension service employees with little or no knowledge of climate change. However, the communication of uncertainty remains a particular challenge. In Kenya a pragmatic approach was adopted in which the project team plans to continue to monitor climate change predictions with a view to bringing them to the community only when uncertainty has reduced. This process of monitoring unfolding climate change predictions is an important aspect of adaptation, and in most contexts will require efforts to institutionalize a system of regular reviews at the district or national level, to ensure that climate information receives continuous attention and is communicated down to communities beyond the life of and independently from any project intervention. Increasing scientists' appreciation of local needs and experiences, and local appreciation of uncertainty and the importance of regularly updating prediction and forecasting information, is also likely to play a pivotal role in ensuring that emerging climate knowledge is better understood at the community level (R. Ewbank, personal communication, 26 September 2008).

Adaptive capacity and social networks

While access to appropriate climate and forecasting information has a particular significance, adaptive capacity has other important components. As summarized in Table 10.1, the case studies document how adaptation projects can: develop leadership (see for example Bangladesh and Niger); provide training on climate change and the skills needed to address its impacts, including literacy (a prerequisite for many adaptation activities and a fundamental development activity, yet inadequately provided for by the authorities in Bangladesh, Niger, Pakistan and Sudan); support experimentation (Sri Lanka, Kenya and Peru); and build the confidence of communities to employ alternative technologies (Kenya, Nepal, Sri Lanka, Pakistan and Peru). Moreover, as Table 10.2 illustrates, social networks play an important role at the heart of adaptive capacity, bringing direct benefits to communities as well as being instrumental in supporting the other elements of adaptive capacity.

The summary in Table 10.2 demonstrates the value of the relationships that form a household or community's social networks. These relationships are the channels for multiple benefits that reach the community either directly or via intermediaries. Most tangibly, resources are secured through networks, but equally important are opportunities for training and information exchange, political engagement and influence in policy issues. The case study chapters

Table 10.2 Social networks in different adaptation contexts

Country	Role of social networks	Lessons
Bangladesh	Community volunteer groups coordinated project CBOs established to build relationships with government and non-government organizations outside community Input into local government disaster preparedness and planning via CBO	Community groups as the basis for participation and capacity building Extending social networks secured resources and enabled collective action in a politically sensitive environment
Kenya	Project established vertical community network links by integrating government personnel as project stakeholders Farmer-to-farmer training and demonstration of technologies Project formed bridge between meteorological office staff and community	Efforts to build understanding between all stakeholders yielded seasonal forecasts communicated in terms of crop and seed types Scaling up opportunities and sustainability of activities built into project through government commitment to project
Niger	Community groups formed to collectively identify problems and manage project interventions Community groups engaged in political lobbying through workshops to develop alternative policy text Leaders of community groups now actively working to influence policy at local and national levels	Marginalization from national networks was a blockage to policy influence on issues such as land rights Interventions to improve literacy enabled interaction with networks Long-term approach built leadership and political engagement within community
Nepal	Informal local groups transformed into CBOs, linking with local government, NGOs and networks at local and district level Forest user groups formed to connect with local government and district forest office to develop sustainable management plans CBO empowered to coordinate work implemented by other agencies Relationships developed with government and non-government service providers	CBO and user group networks provided opportunities to share experiences and gain access to information, training and resources Linking to market networks helped sustain access to resources CBOs help communities maintain links with government service providers during political upheaval Implementing NGO has influence in government climate change network as a result of project
Pakistan	Community organizations created to design and manage project Formation of resource centres where farmers and extension workers meet and information is disseminated	Formal status and co-funding needed before community organization can receive government money Traditional forms of community mobilization can be more effective than external CBO model
Peru	Disaster preparedness committee linked local civil defence into the national system, ensuring that community mobilization is in step with regional and local government Project linked agricultural scientists with farmers	Tradition of community organization facilitated the formation of a new committee

Country	Role of social networks	Lessons
Sri Lanka	Existing paddy farmer group used as main vehicle for project Project designed to bring together community and a wide variety of local government, research institute and non-governmental personnel Farmer-to-farmer learning network created	Even local formal institutions can exclude the poor from their networks Research institute attitude to community revolutionized through relationship building Horizontal networks between communities are effective for disseminating adaptation lessons
Sudan	Strengthened farmers' union became member of the state union, with a voice to government Increased access to meteorological information as a result of links forged through the project	Capacity building and literacy classes were a catalyst in groups becoming effective

also illustrate the importance of social networks for the sustainability of project interventions (for example in Nepal where a secure relationship with government service providers ensured ongoing access during political and institutional upheaval) and scaling up project activities (through linking the communities to government institutions, as in Kenya and Sudan, or providing opportunities to share adaptation experiences). While most of the benefits of social networks accrue to adaptive capacity (enabling communities to create positive changes), access to natural resource institutions such as water or forest management committees can also strengthen resilience by enabling communities to provide feedback on policy proposals, ensuring their voice is heard and thereby preventing damaging changes. Here, the overlap between adaptive capacity and resilience becomes clear: adaptive capacity builds the skills and opportunities to bring about advantageous changes, thereby supporting resilience-strengthening measures that reduce the negative impact of changes as they emerge.

The presence of a focus on social networks in all of the case studies reflects the importance of networks to adaptation. In almost all examples the process of extending networks started with identifying existing, or forming new, community groups to implement project design and management activities. This approach provides an opportunity to build capacity in collective decision making and in many cases resulted in the translation of local informal groups into registered community-based organizations. In many contexts this formalization allows communities to generate vertical connections into the institutions of local and national government, opening up access to resources and information and enabling policy influence. The projects in Kenya and Sudan adopted a different – and effective – approach in linking the community and government as joint stakeholders in the project. Effective policy influence was achieved in Niger through community groups coming together with politicians in a workshop setting, while in Bangladesh the formation of CBOs enabled communities to operate within a difficult political environment in which access to resources is routinely contested.

While vertical contacts to the next level of government are important for community-based adaptation, many of the case studies make use of horizontal networks to support and share adaptation practice. Farmer-to-farmer networks, such as those formed in Sri Lanka and Kenya, enabled project experiences to be taken beyond the project community and helped reduce the perceived risk of change by providing exposure to examples of successful practice. Many projects also built networks with local and national non-governmental and civil society organizations, again bringing opportunities for access to information, learning and resources. The use of horizontal networks to provide a platform for policy lobbying or wider social change to support adaptation was, however, unexplored in the case studies. This points to the omission of systematic network analysis in project design or during the baseline survey, reflecting on the one hand the politically weak situation that many of the communities are in, and on the other a lack of experience in network analysis, mobilization and social organization among the staff in the implementing NGOs.

In Pakistan problems were encountered when it emerged that CBOs needed to generate co-funding before the government would provide financial support. This barrier to effective action highlights the need for a comprehensive analysis that includes the social, political and economic dimensions of potential relationships as part of an adaptation intervention. Analysis of social networks also reveals processes of exclusion and where efforts need to be invested to secure redress. In the Sri Lanka case study, for example, it was noted that the local rice research institute failed to recognize the needs of small-scale producers, yet active relationship building, instigated by the project team, significantly altered the attitude of the research professionals to the benefit of the community. Similarly, bridging the knowledge systems of the local community and meteorological office staff in Kenya enabled modern science to be of value to the poor. Different problems were encountered in Niger, where literacy, remoteness and a nomadic lifestyle were factors in the marginalization of pastoralists from policy processes. Moreover, the experiences of policy engagement in Niger and Kenya both point to the inevitable power dynamics at play in the politics of influence and the challenge of overcoming established interests.

Scaling up community-based adaptation

Scaling up is the crucial step in rolling out project learning to the ever-increasing number of communities vulnerable to climate change. It is a substantial challenge, which needs be addressed through multiple processes. An important element is the replication of the approaches that have been developed by those working with communities. Successful replication has to involve the adoption of good practices, dissemination of relevant knowledge and technologies, and above all, building of capacity among a variety of stakeholders, depending on the political, social and economic context: local NGOs, community leaders, and local and district government officials.

As the preceding chapters have shown, access to information is key to adaptation. Information on climate trends and seasonal weather forecasts will help determine the technologies and livelihood strategies that will be most appropriate. Communities need to be able to combine their own knowledge of what works, with new knowledge in tune with their culture and values. For this reason, if adaptation is to go to scale within a district it is essential to build the capacity of local organizations to manage change, to access information, and to get their voices and concerns heard by local decision makers. For sustainability, links with local government and other stakeholders that hold this knowledge must be built – but for these links to be effective, all stakeholders need to be aware of the significance of climate change and its local impacts on their own work and on the lives of the local population.

Capacity building at the local or district level can be achieved through NGO programmes that link CBOs with local government and other stakeholders (including the local private sector); simultaneously national governments must invest in building the capacity of their district-level staff engaged in natural resource management and agriculture. Establishing district- and national-level networks for sharing experiences among CBOs, NGOs and government departments will be essential for maintaining information flows: expanding social networks to include government actors is thus crucial in scaling up adaptation (as illustrated by the Kenya and Nepal case studies in particular). While formal networks between community and official bodies are valuable, informal networks between communities (as in Sri Lanka) will be important for replication of good practice and the sharing of knowledge and technologies (such as the seeds of resilient crops).

As noted in the Introduction, current estimates of the cost of adaptation vary and do not always consider community-based approaches. While the World Bank is currently engaged in research to produce estimates based on a more detailed assessment than the current figures, this study will focus on national level costs rather than community-based adaptation. The examples in the case studies in this book offer a limited contribution to the currently shallow pool of knowledge of the costs of community-based adaptation. Costs per beneficiary per year varied between £47 (Bangladesh) and £135 (Kenya), with several countries in the region of £60 to £84. The term beneficiary referred to a person in a household receiving training or material support. The costs between projects are not directly comparable as they reflect a different range of activities in each country, albeit with a significant number of common elements (training, capacity building, focus on agriculture and land management, relatively little on high-cost material inputs). Clearly there are also marked cost differentials between implementing programmes in countries with different wage costs (Kenya and Sri Lanka being relatively expensive). It is therefore not possible using this limited data to produce a generalized figure for community-based adaptation. However, focusing on Bangladesh alone provides an example: based on around 60 million extremely poor people living in rural areas vulnerable to climate change, support for adaptation of their livelihoods totals close to £3 billion per year for adaptation interventions. While this is a huge figure, a focus on the cost of project replication fails to describe the full challenge of scaling up community-based adaptation.

Policy, politics and power: the challenge of community empowerment

Discussion of scaling up presupposes an enabling and supportive institutional and policy environment for the implementation of community-based adaptation. However, two issues in particular are significant barriers to the widespread application of the community-based adaptation model. The first is the willingness of governments to spend resources on the poorest section of their population. Vulnerability to climate change is substantially linked to the ability to access and control resources, and to the opportunity and skills to influence decisions that affect livelihoods. As the case studies suggest, governments in developing countries frequently have a poor record of enabling resource access for the poor. However, flows of funds for adaptation will be to national governments, and it will be for the UNFCCC and Adaptation Fund Board to meet the challenge of ensuring that money reaches the most vulnerable (the Board's draft guidelines require priority be given to 'systems, sectors and communities that are particularly vulnerable'; The Adaptation Fund, 2008: 5). Granting funds subject to a partnership agreement with civil society stakeholders or national or international NGOs with experience in implementing adaptation programmes would be a positive step in this regard.

The second issue is whether the national and international institutional and policy context can act as an enabling environment, making it easier for people to improve their livelihoods. Too often the reverse is true and policies are enacted that prevent people from adopting strategies that would help them cope with shocks. An example of this is the widespread promotion of the dominant agricultural model, predicated on the intensive production of a limited number of crops using varieties that often depend on predictable water supply and regular applications of fertilizer (an increasingly costly and fossil-fuel derived commodity). The power of the global seed industry is at play here: 10 companies own 60 per cent of the world's agricultural seed supply, focusing on just four staple crops (ETC Group, 2007). Such agricultural systems are the antithesis of climate resilient and adaptive agriculture, as the recent International Assessment of Agricultural Knowledge, Science and Technology for Development report concludes (IAASTD, 2008). Yet while seed companies have become a dominant force in world agriculture, the importance of seed saving remains and cannot be overstated: financially, in terms of saving annual input costs and reducing the need to rely on credit, but more importantly as a means of protecting biodiversity, ensuring that there will be a gene pool from which to breed climate change-resilient food crops now and in the future. As the Sri Lanka, Kenya and Peru case study chapters demonstrate, farmers' plant breeding activities, drawing on their traditional knowledge, are a flexible strategy for generating new cultivars using different local varieties. It has the added advantage of empowering local farmers and women – meeting two crucial requirements for successful adaptation. Yet global seed companies, with the support of the international intellectual property rights (IPR) regime under the World Intellectual Property Organization and the World Trade Organization, defend their right to patent new varieties and even indigenous varieties, taking control of one of the most valuable resources available to small-scale farmers (Tansey, 2008: 6–14).

The policy and practice of governments at all levels will also need to change if access to information is to be improved. The prevailing model by which information provision and extension operates is heavily centralized and overly bureaucratic, and has proven inefficient and ineffective. A general shift in policy approaches towards liberalization and privatization of services may lead to some groups (particularly poor and remote farmers) receiving no service (Neuchatel Group, 2002; Chapman and Tripp, 2003; Christoplos and Farrington, 2004; Alex et al., 2004). This apparent trend is illustrated in a number of the case studies. An alternative approach, suggested by the experiences described in the Peru chapter, would involve a change in policy by governments. In this model, the widespread building up and development of a community-based extension system could provide an alternative to costly public extension systems, with government departments providing a regulatory role and updating technical skills, and with private agricultural service providers operating as a channel for inputs such as seeds, tools and fertilizers. In the Pakistan and Niger projects, community-based veterinary services proved successful in meeting the needs of rural farmers using a similar approach. The success of these examples reflects the findings of a recently completed Practical Action evaluation of community-based extension programmes in five countries. Five years after the end of funding, more than 50 per cent of the trained extension workers were found to be earning a living delivering information and animal health services within their own communities (Coupe, 2008).

Putting the needs of communities at the heart of local government service is required for scaling up adaptation. This means listening to the voices of the small farmer producers' groups. It means allocating resources to support knowledge exchange between communities through exchange visits and farmer-to-farmer extension. It involves direct partnership between government, local NGOs and CBOs in project design and implementation. It requires national and international research institutes to take up the research needs of smallholders on food production, processing and marketing. Evidence from the case studies suggests that agricultural research institutes frequently do not involve farmers as active partners in their research and do not take farmers' problems as the starting point for their research programmes. Effective adaptation will require a change in practice and priorities on their part, and the widespread dissemination of relevant findings, in accessible language, designed to reach communities. Research institutes will need to commit to run training courses for local NGOs and community groups to update their skills, reflecting climate change as an ongoing challenge that will require the continuous review of appropriate technologies for adaptation strategies.

The connections between public policy and adaptation are therefore clear. The usefulness of climate and forecasting information depends on public investment in scientific research, guided by policies that understand the needs of small-scale farmers. Rural communities are dependent on natural resources whose management is frequently the responsibility of public institutions. Publicly funded extension services have the potential to distribute resources and support knowledge sharing in marginal environments (which can in turn support micro-entrepreneurial community-based extension systems). Nationally and internationally funded agricultural research institutes must turn their focus to the needs of the most vulnerable. In short, adaptation can be subject to a favourable, enabling policy environment that is reflective of the interests and needs of rural communities. Yet poor communities are frequently marginalized from policy processes and by regulatory controls, and their interests are unrepresented and overlooked in decision making. The Niger case study reflects this. The process through which the pastoralist community should be able to secure use rights over their lands is complex and poorly enforced, leaving the community's investment in land regeneration vulnerable to transitory herders (a phenomenon that is itself partly due to the effects of national policy). A more favourable environment would simplify and enforce the recognition of use rights for pastoralist communities, while an enabling environment might seek to promote, support and provide an appropriate legal framework for community-based natural resource management. A positive example is provided by the decentralized local planning process described in the Peru chapter, through which the community was able to have its needs included in local government budgets.

For adaptation, local decision making offers benefits in terms of resilience and adaptive capacity by being responsive to local knowledge of environmental risks and opportunities. At the least, devolved decision making offers resilience: for example, where forest and water user groups are able to reduce the chance of poor decision making. However, at its best, local decision making includes an active role for informed communities, enabling them to act on the basis of their self-defined best interest identified through equitable processes of engagement. If achievable, this level of community empowerment fulfils a significant component of adaptive capacity. The ultimate enabling environment is therefore one in which support is given at the local level to secure a 'political, social and cultural environment that encourages freedom of thought and expression, and stimulates inquiry and debate' (Twigg, 2007: 26). While efforts at the intergovernmental level may seek to ensure that financial resources reach those most in need, a substantial challenge remains for those engaged in community-based adaptation: are they prepared to engage in the processes of change necessary to secure, or at least work towards, an enabling policy environment? This is a complex and context-specific problem: few poor communities are in areas where there are institutions ready to adopt the principles of local accountability, and in many circumstances a radical transformation may be unrealistic. Yet there is a spectrum of outcomes through which decision making becomes increasingly localized, including networks of user groups linked to institutions, embedded processes of consultation, and local participation in decision making. Advocacy for change can be supported through the sorts of process illustrated by the examples in this book: strengthened local networks, enhanced relationships between community-based networks and institutions, and facilitated policy discussions. International decision making, while not considered in the case studies, clearly has ramifications for the ability to adapt at the local level (for example, through seed IPRs). Lobbying and influence is difficult to achieve in these forums and requires training and resources; here, international NGOs should look to support activists from community networks.

The community-based approach therefore has the potential to help deliver an enabling policy environment through established mechanisms: enhancing social networks and retaining a focus on the processes of engagement – participation, equality and respect – that respond to local interests. Its methodology has at its root a focus on engagement with communities that enables adaptation projects to be responsive to the local context. The challenge of establishing an enabling policy environment, and thus of scaling up, is to take the principles and processes that are inherent in community-based practice and embed them into the policies that govern administrative functions, decision making and accountability.

Cost estimates based on replicating project interventions suggest that adaptation carries enormous financial implications. Yet, as this conclusion has argued, it is a lesson of the case studies that it is the transformation of governance and policy frameworks that is necessary: networks of support for community groups and farmers' organizations with information and resources guided by responsive district government, appropriate research and effective extension. Here, the principal cost is political; the cost per beneficiary is reduced through the economies of scale that social networks generate (as in Sri Lanka, for example, where farmer-to-farmer links enabled project learning to be disseminated outside the community and transformed the focus of the local research institute). Money will still be needed for inputs, research, training and improved meteorological support, but it is essential to recognize that, without addressing the governance issues at the heart of exclusion, project-based adaptation will be left to meet the challenge of climate change – at enormous financial cost. Ultimately the price will be in human lives, as it is infeasible that NGOs running interventions replicating best adaptation practice could reach even the 60 million people in Bangladesh who are vulnerable to climate change.

References

Abdel Atti, H. and Elhassan, H. (2003) *Environmental Strategies to Increase Human Resilience to Climate Change, Lessons for Eastern and North Africa*, Khor Arba'at Rehabilitation Report conducted under AIACC-AF14.

Adger, W.N. (2003) 'Social capital, collective action, and adaptation to climate change', *Economic Geography* 79(4): 387–404.

Alex, G., Bayerlee, D., Helene-Colion, M. and Rivera, W. (2004) *Extension and Rural Development, Converging Views on Institutional Approaches?* [online], World Bank Agriculture and Rural Development Discussion Paper 4, available from: http://www-wds.worldbank.org/servlet/main?menuPK=64187510&pagePK=64193027&piPK=64187937&theSitePK=523679&entityID=000160016_20050706164415 [accessed 5 December 2008].

Balstad, R. (2008) 'Adapting to an uncertain climate on the Great Plains: testing hypotheses on historical populations', in *Living With Climate Change: Are There Limits to Adaptation?* pp. 166–174, Conference Proceedings 7–8 February 2008, Royal Geographical Society, London.

Brooks, N. (2003) *Vulnerability, Risk and Adaptation: A Conceptual Framework*, Working Paper 38, Tyndall Centre for Climate Change Research, Norwich, UK.

Burton, I., Huq, S., Lim, B., Pilifosova, O. and Schipper, E.L. (2002) 'From impacts assessment to adaptation priorities: the shaping of adaptation policy', *Climate Policy* 2(2–3): 145–159.

Cash, D.W., Borck, J.C. and Patt, A.G. (2006) 'Countering the loading-dock approach to linking science and decision making', *Science, Technology and Human Values* 31(3): 1–30.

Chapin, F.S., Lovecraft, A.L., Zavaleta, E.S., Nelson, J., Robards, M.D., Kofinas, G.P., Trainor, S.F., Peterson, G.D., Huntington, H.P. and Naylor, R.L. (2006) 'Policy strategies to address sustainability of Alaskan boreal forests in response to a directionally changing climate', *Proceedings of the National Academy of Sciences of the United States of America* 103(45): 16,637–16,643.

Chapman, R. and Tripp, R. (2003) 'Changing incentives for agricultural extension – a review of privatised extension in practice', *ODI Agren Newsletter* 132, July 2003, ODI, London.

Christensen, J.H., Hewitson, B., Busuioc, A., Chen, A., Gao, X., Held, I., Jones, R., Kolli, R.K., Kwon, W.-T., Laprise, R., Magaña Rueda, V., Mearns, L., Menéndez, C.G., Räisänen, J., Rinke, A., Sarr, A. and Whetton, P. (2007) 'Regional climate projections', in S. Solomon, D. Qin, M. Manning, Z. Chen, M. Marquis, K.B. Averyt, M. Tignor and H.L. Miller (eds), *Climate Change 2007: The Physical Science Basis. Contribution of Working Group I to the Fourth Assessment Report of the Intergovernmental Panel on Climate Change*, pp. 848–940, Cambridge University Press, Cambridge, UK.

Christoplos, I. and Farrington, J. (2004) *Poverty, Vulnerability and Agricultural Extension: Policy Reforms in a Globalising World*, Oxford University Press, Delhi.

Cleaver, F. (2001) 'Institutions, agency and the limitations of participatory approaches to development', in B. Cooke and U. Kothari (eds), *Participation: The New Tyranny?* Zed Books, London.

Coupe, S. (2008) 'Preliminary Review of Practical Action's Community Based Extension Experience', Practical Action unpublished report.

Coupe, S.A., Hellin, J., Masendeke, A. and Rusike, E. (2005) *A Farmers' Jury: The Future of Smallholder Agriculture in Zimbabwe*, ITDG Publishing, Rugby.

Cross, C.L. and Parker, A. (2004) *The Hidden Power of Social Networks*, Harvard Business Press, Boston, Massachusetts.

Desai, S., Hulme, M., Lempert, R. and Pielke, R. (2008) 'Climate prediction: a limit to adaptation', in *Living With Climate Change: Are There Limits to Adaptation?* pp.49–57, Conference Proceedings 7–8 February 2008, Royal Geographical Society, London.

Ensor, J. (2005) 'Linking rights and culture: implications for rights-based approaches' in P. Gready and J. Ensor (eds), *Reinventing Development? Translating Rights-Based Approaches from Theory into Practice*, pp. 254–277, Zed Books, London.

Ensor, J.E. and Berger, R. (2009, forthcoming) 'Community Based Adaptation and Culture in Theory and Practice', in W. N. Adger, I. Lorenzoni and K. O'Brien (eds), *Adapting to Climate Change: Thresholds, Values, Governance*, Cambridge University Press, Cambridge.

Eriksen, S.E.H., Klien, R.J.T., Ulsrud, K., Naess, L.O. and O'Brien, K. (2007) *Climate Change and Poverty Reduction: Key Interactions and Critical Measures*, Global Environmental Change and Human Security (GECHS) Report 1, [online], available from: http://www.norad.no/default.asp?V_ITEM_ID=10502 [accessed 5 December 2008].

Eriksson, M. (2006) *Climate Change and its Implications for Human Health in the Himalaya*, Sustainable Mountain Development 50, ICIMOD, Kathmandu.

ETC Group (2007) *The World's Top 10 Seed Companies in 2006* [online], ETC Group, available from: www.etcgroup.org/upload/publication/pdf_file/615 [accessed 20 October 2008].

Food and Agriculture Organisation (FAO) (2007) *Adaptation to Climate Change in Agriculture, Forestry and Fisheries: Perspective, Framework and Priorities*, FAO, Rome.

Gine, X., Townsend, R. and Vickery, J. (2007) *Patterns of rainfall insurance participation in rural India* [online], New York Fed Staff Reports 302, available from: http://www.newyorkfed.org/research/economists/vickery/WBER_India_Insurance_August_10_2007.pdf [accessed 5 December 2008].

Hare, B. (2008) *The Science of Climate Change*, Breaking the Climate Deadlock Briefing Paper, The Climate Group [online], available from: http://www.theclimategroup.org/assets/resources/Science_of_Climate_Change.pdf [accessed 5 December 2008].

Harrison, M., Troccoli, A., Anderson, D.L.T. and Mason, S.J. (2007a) 'Introduction', in A. Troccoli, M. Harrison, D.L.T. Anderson and S.J. Mason (eds), *Seasonal Climate: Forecasting and Managing Risk*, pp. 3–12, Springer Academic Publishers, London.

Harrison, M., Troccoli, A., Anderson, D.L.T., Mason, S.J., Coughlan, M. and Williams, J.B. (2007b) 'A way forward for seasonal climate services', in A. Troccoli, M. Harrison, D.L.T. Anderson and S.J. Mason (eds), *Seasonal Climate: Forecasting and Managing Risk*, pp. 413–425, Springer Academic Publishers, London.

Hawe, P., Webster, C. and Shiell, A. (2004) 'A glossary of terms for navigating the field of social network analysis', *Journal of Epidemiology and Community Health* 58: 971–975.

Heffernan, O. (2008) 'They say they want a revolution', *Nature* 453: 268–269.

International Assessment of Agricultural Knowledge, Science and Technology for Development (IAASTD) (2008) *Executive summary of the synthesis report* [online], available from: http://www.agassessment.org/docs/SR_Exec_Sum_280508_English.pdf [accessed 20 October 2008].

IPCC (1988) *Mandate* [online], IPCC, available from: http://www.ipcc.ch/about/index.htm [accessed 4 August 2008].

IPCC (2007) 'Summary for policymakers', in M.L. Parry, O.F. Canziani, J.P. Palutikof, P.J. van der Linden and C.E. Hanson (eds), *Climate Change 2007: Impacts, Adaptation and Vulnerability. Contribution of Working Group II to the Fourth Assessment Report of the Intergovernmental Panel on Climate Change*, pp. 7–22, Cambridge University Press, Cambridge, UK.

Jennings, T.L. (2008) 'Adaptation to climate change in the UK: a case study of the Boscastle harbour flood disaster', in *Living With Climate Change: Are There Limits to Adaptation?* Conference Proceedings 7–8 February 2008, pp. 138–147, Royal Geographical Society, London.

Kelly, P.M. and Adger, W.N. (2000) 'Theory and practice in assessing vulnerability to climate change and assessing adaptation', *Climate Change* 47: 325–352.

Kymlicka, W. (1989) 'Liberalism, individualism and minority rights', in A.C. Hutchinson and L.J.M. Green (eds), *Law and the Community*, pp. 181–204, Carswell, Toronto.

Kymlicka, W. (1995) *Multicultural Citizenship*, Clarendon Press, Oxford, UK.

Lemons, J. (2007) 'Climate change: the normative dimensions of IPCC's approach to scientific uncertainty', ClimateEthics.org [online], available from: http://climateethics.org/?p=25 [accessed 4 August 2008].

Meehl, G.A., Stocker, T.F., Collins, W.D., Friedlingstein, P., Gaye, A.T., Gregory, J.M., Kitoh, A., Knutti, R., Murphy, J.M., Noda, A., Raper, S.C.B., Watterson, I.G., Weaver A.J. and Zhao, Z.-C. (2007) 'Global climate projections', in S. Solomon, D. Qin, M. Manning, Z. Chen, M. Marquis, K.B. Averyt, M. Tignor and H.L. Miller (eds), *Climate Change 2007: The Physical Science Basis. Contribution of Working Group I to the Fourth Assessment Report of the Intergovernmental Panel on Climate Change*, pp. 748–845, Cambridge University Press, Cambridge, UK.

Millennium Ecosystem Assessment (2005) *Ecosystems and Human Well-being: Synthesis*, Island Press, Washington, DC.

Möhner, A. and Klein, R.J.T. (2007) *The Global Environment Facility: funding for adaptation or adapting to funds?* Stockholm Environment Institute.

Moser, S.C. (2008) 'Whether our levers are long enough and the fulcrum strong? Exploring the soft underbelly of adaptation decisions and actions', in *Living With Climate Change: Are There Limits to Adaptation?* Conference

Proceedings 7–8 February 2008, pp. 176–193, Royal Geographical Society, London.

Müller, B. (2008) *International Adaptation Finance: The Need for an Innovative and Strategic Approach*, Working Paper EV 42, Oxford Institute for Energy Studies, Oxford, UK.

Neuchatel Group (2002) *Common Framework on Financing Agricultural and Rural Extension*, Neuchatel Group, Lindau, Switzerland.

O'Brien, K., Eriksen, S., Schjolden, A. and Nygaard, L. (2004) *What's in a Word? Conflicting Interpretations of Vulnerability in Climate Change Research*, Working Paper 04, CICERO, Oslo, Norway.

Orindi, V.A. and Murray, L.A. (2005) 'Adapting to climate change in East Africa: a strategic approach', *IIED Gatekeeper Series* 117, IIED, London.

Oxfam (2007) *Adapting to Climate Change*, Briefing Paper 104, Oxfam, Oxford, UK.

Pachauri, R. (2006) 'Avoiding dangerous climate change', in H.J. Schellnhuber (ed.), *Avoiding Dangerous Climate Change*, pp. 3–6, Cambridge University Press, Cambridge, UK.

Palmer, T.N., Doblas-Reyes, F.J., Weisheimer, A. and Rodwell, M.J. (2008) 'Toward seamless prediction. Calibration of climate change projections using seasonal forecasts', *Bulletin of the American Meteorological Society* 89: 459–470.

Patt, A. (2005) 'Effects of seasonal climate forecasts and participatory workshops among subsistence farmers in Zimbabwe', *Proceedings of the National Academy of Sciences of the United States of America* 102(35): 12,623–12,628.

Patt, A. (2008) 'How does using seasonal forecasts build adaptive capacity?' in *Living With Climate Change: Are There Limits to Adaptation?* Conference Proceedings 7–8 February 2008, pp. 62–67, Royal Geographical Society, London.

Pearce, F. (2008a) 'Poor forecasting undermines climate debate', *New Scientist* 2654: 8–9.

Pearce, F. (2008b) 'Climate scientists call for their own "Manhattan Project"', NewScientist.com news service, 7 May 2008 [online], http://environment.newscientist.com/channel/earth/climate-change/dn13855-climate-scientists-call-for-their-own-manhattan-project.html [accessed 4 August 2008].

Pelling, M. and High, C. (2005) 'Understanding adaptation: what can social capital offer assessments of adaptive capacity?' *Global Environmental Change A* 15(4): 308–319.

Peterson, G. (2000) 'Political ecology and ecological resilience: an integration of human and ecological dynamics', *Ecological Economics* 35: 323–336.

Peterson, G., De Leo, G.A., Hellmann, J.J., Janssen, M.A., Kinzig, A., Malcolm, J.R., O'Brien, K.L., Pope, S.E., Rothman, D.S., Shevliakova, E. and Tinch, R.R.T. (1997) 'Uncertainty, climate change, and adaptive management', *Conservation Ecology* 1(2): 4 [online], http://www.consecol.org/vol1/iss2/art4/ [accessed 11 December 2008].

Phillips, J. (2003) 'Determinants of forecast use among communal farmers in Zimbabwe', in K. O'Brien and C. Vogel (eds), *Coping with Climate Variability: The Use of Seasonal Climate Forecasts in Southern Africa*, pp. 110–128, Ashgate, Aldershot, UK.

Raz, J. (1988) *The Morality of Freedom*, Clarendon Press, Oxford, UK.
Smit, B. and Wandel, J. (2006) 'Adaptation, adaptive capacity and vulnerability', *Global Environmental Change* 16: 282–292.
Stavenhagen, R. (1998) 'Cultural rights: a social science perspective', in H. Niec (ed.), *Cultural Rights and Wrongs*, UNESCO Publishing, Paris.
Tansey, G. (2008) 'Food, farming and global rules', in G. Tansey and T. Rajotte (eds), *The Future Control of Food*, pp. 3–26, Earthscan, London.
The Adaptation Fund (2008) *Draft strategic priorities, policies and guidelines of the Adaptation Fund*, The Adaptation Fund Board Third Meeting, Agenda Item 6, AFB/B.3/9.
Twigg, J. (2007) *Characteristics of a Disaster-resilient Community*. Benfield UCL Hazard Research Centre, London.
UNDP (2007) *Human Development Report 2007/2008 Fighting Climate Change: Human Solidarity in a Divided World*, UNDP New York.
Wasserman, S. and Faust, K. (1994) *Social Network Analysis: Methods and Applications*. Cambridge University Press, Cambridge, UK.
World Bank (2006a) *Clean Energy and Development: Towards an Investment Framework*, World Bank Environmentally and Socially Sustainable Development and Infrastructure Vice Presidencies, World Bank, Washington DC.
World Bank (2006b) *An Investment Framework for Clean Energy and Development: A Progress Report*, DC2006–0012, World Bank, Washington, DC.
World Climate Research Programme (2008a) *World Modelling Summit for Climate Prediction, Reading, UK, 6–9 May 2008* [online], available from: http://wcrp.ipsl.jussieu.fr/Workshops/ModellingSummit/Documents/WMS_prospectus_20080306.pdf [accessed 4 August 2008].
World Climate Research Programme (2008b) *The Climate Prediction Project*. [online], available from: http://wcrp.ipsl.jussieu.fr/Workshops/ModellingSummit/Documents/FinalSummitStat_6_6.pdf [accessed 4 August 2008].

Index

agriculture *see* farming
AIACC *see* Assessments of Impacts and Adaptation to Climate Change
alpaca rearing 148–50, 152, 156–7, 160
analytically-based reasoning 19–20
Andes mountains *see* Peru
animal health technologies 157
anticipated adaptation 15–16
arid climates 101–14
art, awareness-raising 46, 50
Assessments of Impacts and Adaptation to Climate Change (AIACC) 131–2, 139
awareness
    Bangladesh 41–6, 49, 50–1
    Kenya 105–7, 111
    Nepal 60–1
    Niger 118, 119, 121, 126
    Pakistan 72, 73–4, 75–9, 83–4
    Peru 149–53, 161
    Sri Lanka 89, 91–2, 96, 97–8
    Sudan 133
ayllus/ayni social organizations, Peru 155

Bangladesh 39–45
banners, awareness-raising 78
bari agricultural land 56–7
barriers against flooding 64
    see also embankments
biodiversity 159
bird behaviour 153
bonding capital 21
'boundary organizations' 23
brochures, awareness-raising 106, 108, 110
bulletins, awareness-raising 42–3, 152

caged fish cultivation 47
calendars, awareness-raising 106, 109–10
camel rearing 77
    see also alpaca rearing
campaigns, awareness-raising 78–9
capacity building 17–25, 26, 28, 30, 32–3
    Bangladesh 49
    Peru 154–5
    Sudan 134, 140, 142
capital 21
    see also economics
CCBs (Citizen Community Boards) 74
celebrations, awareness-raising 46, 77–8
cement well construction 122
char islands 40
    see also Bangladesh
Citizen Community Boards (CCBs) 74
clarity 6, 29–33, 98
cloud reading 153
coastal climates 87–100, 131–45
community-based adaptation 1–38
    Bangladesh 41–9
    Kenya 102, 103–11
    Nepal 58–68
    Niger 117–24
    Pakistan 73–82

Peru 149–59
Sri Lanka 87, 89–97
Sudan 134–43
computing power 10–11
concrete well construction 122
continuous hazards 13, 15
   see also hazards
contour bunds 137
cooking stoves 47, 51
Copenhagen agreement 5–6
coping strategies 18–19, 24–5
   Bangladesh 46
   Kenya 111
   Niger 118–19
   Peru 149
   Sri Lanka 97
   Sudan 135, 139
   see also resilience
credit systems 138
   see also economics
crop growth *see* farming
culture 33–6, 45, 77, 125

dams
   Kenya 110–13
   Nepal 64, 67
   Sudan 143
date palms 137
de-stocking 125–6
debates, awareness-raising 45, 50
deforestation 55–8, 65–6, 122
demonstration sites 109
desertification 71–86
   Niger 116
   Peru 151–2
development 1–3, 5
   Bangladesh 46–8, 49
   Kenya 108–10
   Nepal 61–4
   Niger 121–4, 127–8
   Pakistan 72, 79–81, 82
   Peru 156–9
   Sri Lanka 93, 94–6
   Sudan 132, 134–5, 137
   see also sustainability

disaster relief 1–2
   Bangladesh 43–4
   Kenya 107–8
   Niger 117, 127
   Sri Lanka 93, 96, 98
   Sudan 134
discrete hazards 13, 15
discussions
   awareness-raising 77, 139
   see also focus groups
documentaries, awareness-raising 78
drought 3, 24–5
   Bangladesh 39–40
   Kenya 101–14
   Niger 115–17, 124–6, 127
   Peru 150
   Sri Lanka 97
   Sudan 133–4, 135–6, 142
   see also precipitation
duck rearing 47
dykes 123, 126, 128–9

'early warning committees' 44
earth-moving machinery 136
economics 2–6, 14, 17–20, 21
   Bangladesh 49, 51–2
   Kenya 102, 103–4, 110–11, 112
   Nepal 56–8, 60, 61, 66–7
   Niger 118, 123–4, 126, 128–9
   Pakistan 76, 81–3
   Peru 153, 154, 159
   Sri Lanka 89–90, 91, 93, 95–7
   Sudan 132, 136–7, 138–9, 141–3
   see also poverty
education 27
   Bangladesh 43–4, 50–1
   Nepal 57, 60, 61, 66
   Niger 120, 121, 125–6, 127–8
   Pakistan 75
   Peru 154
   Sri Lanka 87–8, 98
   Sudan 135, 138–40, 141
El Niño Southern Oscillation (ENSO) 8–9, 147–8
elevated structures 48, 51

elite populations (Bangladesh) 40
embankments 51–2, 136
    see also barriers against flooding
end-point vulnerability 14–16
ENSO *see* El Niño Southern Oscillation
equity enhancement 137
erosion 39–45
    Pakistan 72
    Sri Lanka 95
essay competitions 46, 50
experience-based knowledge 19–20, 91
extinctions 74, 116–17
extreme weather 1–4, 6, 12, 151
    Nepal 55–6, 58–9
    Niger 119
    Peru 147–62
    Sri Lanka 96–7
    see also drought

face-to-face interviews 103–4
facilitators 125
faith 125
farming 2, 12, 24–5, 115–30
    awareness festivals 79
    Bangladesh 40, 42, 47–8, 49, 50–1
    community-based adaptation 2–3, 14, 16, 18–20, 22–5
    de-stocking 125–6
    Kenya 102, 104–13
    Nepal 55–6, 59, 61–2, 64–8
    Niger 116–17, 119, 123, 125–9
    Pakistan 71–2, 79, 80–1, 84–5
    Peru 148, 149–52, 154–5, 157–61
    Sri Lanka 87–100
    Sudan 132–5, 136–7, 138–40, 142–3
fast onset events 12
first generation adaptation 16
fishing *see* farming
fixation 118, 120, 122–5, 127
flash floods 55–69

see also flooding
floating gardens 47, 51
flood-proof housing 47
flooding 14
    Bangladesh 39–45
    Nepal 55–69
    Sri Lanka 88, 90, 93
    Sudan 133
floodplain adaptation 71–86
flowering plants 153
focus groups, awareness-raising 89–90, 91–2
fodder banks 124, 129, 160
food security
    Nepal 65
    Niger 123–4, 129
    Peru 154
    Sudan 140, 142
forecasts 8–9, 11–13, 23–4, 26, 29
    Bangladesh 50–1
    Kenya 106, 108
    Niger 119
    Peru 153
    Sri Lanka 92
    see also predictability
forest conservation 65
fox calls 153
fruit production
    Nepal 63, 66
    Pakistan 80
    see also farming

gardens 47, 51, 112, 137, 140
generation-based adaptation 16, 92
goat breeding 48, 64
    see also farming
government *see* policy
grafting 80, 81–2
grain banks 124, 129
grassland improvement 156–7
greenhouse gas emissions 4, 5, 6–9, 107, 111
group discussions 139

hand dug wells 121–2

hazards 13–17, 28, 30–2, 58–9
health technologies 157
herbaceous plants 116, 123
  see also reforestation
herbal medicines 157, 160
Himalayas see Nepal
home gardens 137, 140
homestead vegetable gardening 47
household adaptation see
  community-based adaptation

income see economics
Intergovernmental Panel on Climate
  Change (IPCC) 1–2, 7–9, 131–2
international events/intervention
  4–6, 10–11, 77–8
International Union for
  Conservation of Nature (IUCN)
  89–90
interviews
  Kenya 103–4
  Peru 149
  Sudan 132, 136, 139
IPCC see Intergovernmental Panel
  on Climate Change
irrigation
  Nepal 55–6, 59, 63, 65, 67
  Pakistan 71–2
  Peru 152, 158, 160
  Sri Lanka 87–8, 94
  Sudan 132–3, 137
IUCN see International Union for
  Conservation of Nature

jaal (tree) 77
Jeunesse En Mission Entraide et
  Développement (JEMED) 117–20,
  123–6, 128–9

Kahani Raat (story nights) 76–7
Kamayoq training 154–5
Kenya 101–14
  awareness 105–7, 111
  community-based adaptation
    102, 103–11

development 108–10
drought 101–14
economics 102, 103–4, 110–11,
  112
farming 102, 104–13
knowledge 105–6, 111
livelihoods 103, 105, 110–11
natural resources 102
sustainability 101–3, 107
water management 104
khet agricultural land 56–7
Khor Arba'at region 132–4
  see also Sudan
Kissan resource centres 79
kitchen gardens 112
knowledge 2–3, 6–7, 11, 20, 24–5,
  26–33
  Bangladesh 41–3, 49
  Kenya 105–6, 111
  modern systems 41–2
  Niger 118–19, 125
  Pakistan 73–4, 82–3
  Peru 149–53, 159, 160–1
  Sri Lanka 89, 91–2, 95, 97–8
  Sudan 134, 135–6, 140, 141–3
  traditional systems 41–2, 160

'lack of blessing' concepts 119
land tenure 121, 140
landslides 58–9
language 78, 91
latrines 48, 51
literacy 27
  Niger 128
  Peru 160
  Sudan 135, 136, 138, 139–40
literature reviews 139
Litha forecasting 92
livelihoods 2, 12, 13–15, 18–19,
  27–9, 31, 34
  Bangladesh 39–41, 49, 50, 51
  Kenya 103, 105, 110–11
  Nepal 55–9
  Niger 116, 117, 123–4, 127
  Pakistan 72–4, 78–81, 82, 84–5

Peru 148, 161
Sri Lanka 89–90, 91, 93
Sudan 134, 139–40, 142
livestock *see* farming
Lok Sath (people's parliament) 76

Maha monsoon climate 88
mapping risk/resources
    Peru 149–50
    Sri Lanka 89–92
marketing 136–7
    see also economics
masonry 124
meetings 77, 106, 139, 141
meteorological data
    Bangladesh 39–40
    Kenya 105–6
    Peru 149
    Sudan 136
migration 133, 136
milk production 116–17, 119, 125
modern knowledge systems 41–2
monsoon climate
    Bangladesh 40
    Nepal 56, 59, 62
    Sri Lanka 88, 95
mud well construction 122

National Federation for Conservation of Traditional Seeds and Agricultural Resources (NFCTSAR) 93–4
natural medicines 157, 160
natural resources 2–3, 17
    Bangladesh 41, 49
    Kenya 102
    Nepal 56
    Niger 122–3, 128
    Peru 149–50, 154
    Sri Lanka 88, 90, 95, 98
    Sudan 133, 134–5
Nepal 55–69
    awareness 60–1
    community-based adaptation 58–68
    development 61–4
    economics 56–8, 60, 61, 66–7
    education 57, 60, 61, 66
    farming 55–6, 59, 61–2, 64–8
    irrigation 55–6, 59, 63, 65, 67
    livelihoods 55–9
    poverty 55, 66
    precipitation 58–9, 62
    social networks 60
    water management 61–2, 65
    workshops 61, 65–6
networking 21, 107–8
    see also social networks
newsletters 106
newspapers 42, 105, 106
NFCTSAR (National Federation for Conservation of Traditional Seeds and Agricultural Resources) 93–4
Niger 115–30
    awareness 118, 119, 121, 126
    community-based adaptation 117–24
    development 121–4, 127–8
    drought 115–17, 124–6, 127
    economics 118, 123–4, 126, 128–9
    education 120, 121, 125–6, 127–8
    farming 116–17, 119, 123, 125–9
    knowledge 118–19, 125
    livelihoods 116, 117, 123–4, 127
    natural resources 122–3, 128
    policy 120, 128
    precipitation 115–16, 117, 118–19
    social networks 120–1
    technology 121–4, 127, 129
    water management 118, 121–2
no-regrets strategies 16–17
nomadic living *see* Niger
non-governmental organizations (NGOs) 19, 23, 25, 27
    Bangladesh 43–4, 51

Nepal 58, 61, 66
Niger 120, 125–6
Pakistan 82
Sri Lanka 96, 98
nurseries 47, 95–6, 98

observations, awareness-raising 42, 139
onset events 12

paddy farming 87–100
Pakistan 71–86
    awareness 72, 73–4, 75–9, 83–4
    community-based adaptation 73–82
    development 72, 79–81, 82
    economics 76, 81–3
    education 75
    farming 71–2, 79, 80–1, 84–5
    knowledge 73–4, 82–3
    livelihoods 72–4, 78–81, 82, 84–5
    policy 72–3, 75–6
    poverty 71, 73–82
    social networks 74–5, 83
    technology 79–81, 84–5
paravets 128
participatory rural appraisal (PRA) 46, 118
pastoralism 115–30
pasture improvement 156–7
people's parliament 76
Peru 147–62
    awareness 149–53, 161
    capacity building 154–5
    community-based adaptation 149–59
    development 156–9
    drought 150
    economics 153, 154, 159
    farming 148, 149–52, 154–5, 157–61
    irrigation 152, 158, 160
    knowledge 149–53, 159, 160–1
    livelihoods 148, 161

    precipitation 147–8, 151–2
    social networks 154–5, 159–60
    sustainability 154, 156–7
    technology 156–9, 160
    water management 151, 156, 158
policy 7, 17–18, 21–2, 27–8
    Bangladesh 52
    Kenya 105, 107–8, 112–13
    Nepal 61
    Niger 120, 128
    Pakistan 72–3, 75–6
    Sri Lanka 98
portable cooking stoves 47
poverty 2–5, 20, 25
    Bangladesh 39–40, 46, 51
    Kenya 105
    Nepal 55, 66
    Niger 118, 128
    Pakistan 71, 73–82
    Sri Lanka 93
    Sudan 133–5, 136
    see also economics; Rural Development Policy Institute
power, community-based adaptation 24–5
PRA *see* participatory rural appraisal
precipitation 8–11, 12, 31–2
    Bangladesh 39–40, 42, 50–1
    Nepal 58–9, 62
    Niger 115–16, 117, 118–19
    Pakistan 72, 74
    Peru 147–8, 151–2
    Sri Lanka 88, 96
    Sudan 132–3
    see also drought
predictability 6–13, 15, 20
    Kenya 106
    Peru 151–3
    Sri Lanka 92
    Sudan 135, 136
    see also forecasts
probability-based forecasts 106
publicity materials, awareness-raising 45, 78

pumping stations 121–2, 143

quantitative impact indicators 126–7
questionnaires 103–5, 139

radio 42, 84, 106–7, 119, 152
rainfall *see* precipitation
raised flood-proof housing 47
rallies, awareness-raising 46
RDPI *see* Rural Development Policy Institute
Red Sea coastal belt *see* Sudan
reforestation 119, 122–3
   see also reseeding; tree plantations
regeneration 122–3, 126–7, 128
regional climate projections 10–11
religion *see* faith
reseeding 156
   see also reforestation; tree plantations
resilience 17–25, 26, 28, 30, 32–3
   Kenya 102, 105
   Nepal 62, 65
   Peru 156
   Sudan 131–2, 134
   see also coping strategies
resources *see* economics
restocking 129
rickshaw adverts 46
risk assessment/management 7–8
   Peru 155
   Sri Lanka 89–92
   see also predictability; uncertainty; vulnerability
rural communities *see* poverty
Rural Development Policy Institute (RDPI) 72–5, 77–8, 84, 85
Rut (awareness-raising publications) 78

salinity 87–100
Salvadora persica (tree) 77
season wells 121

seasonal forecasts 8–9, 12–13, 23–4, 50–1
seawater *see* salinity
second generation adaptation 16
seeds *see* farming
semi-arid climates 101–14, 115–30
semi-nomadic living *see* Niger
slow onset events 12
social adaptation *see* community-based adaptation
social networks 20–5, 26–8, 33
   Bangladesh 43–4
   Nepal 60
   Niger 120–1
   Pakistan 74–5, 83
   Peru 154–5, 159–60
   Sri Lanka 93–4
   Sudan 136–7, 141–2
soils
   degradation 116–17
   fertility 156
   reclamation 135
SOS Sahel *see* Sudan
species loss 74, 116–17
starting-point vulnerability 14–17, 29, 31, 33
story nights 76–7
students *see* education
Sudan 131–45
   awareness 133
   capacity building 134, 140, 142
   community-based adaptation 134–43
   development 132, 134–5, 137
   drought 133–4, 135–6, 142
   economics 132, 136–7, 138–9, 141–3
   education 135, 138–40, 141
   farming 132–5, 136–7, 138–40, 142–3
   irrigation 132–3, 137
   knowledge 134, 135–6, 140, 141–3
   livelihoods 134, 139–40, 142

natural resources 133, 134–5
poverty 133–5, 136
precipitation 132–3
social networks 136–7, 141–2
sustainability 132, 141, 142–3
technology 137, 140
water management 131–45
sunlight 151
surface water 117
sustainability
   Kenya 101–3, 107
   Niger 118, 124
   Peru 154, 156–7
   Sudan 132, 141, 142–3
   see also development

Tamasheq communities *see* Niger
technology
   Bangladesh 46–8, 49, 51–2
   Kenya 108–10
   Nepal 61–6
   Niger 121–4, 127, 129
   Pakistan 79–81, 84–5
   Peru 156–9, 160
   Sri Lanka 94–6
   Sudan 137, 140
television 42, 60–1, 84, 105, 107
temperature 4, 8–12
   Bangladesh 39–40
   Nepal 55, 58
   Niger 115
   Peru 148–9
   Sri Lanka 88
timber smuggling 65
traditional knowledge systems 41–2, 160
training
   Bangladesh 44–5, 50–1
   Kenya 108–9, 112
   Nepal 60–2, 64–6
   Niger 128
   Pakistan 81, 83–5
   Peru 154–5, 156, 160
   Sudan 135, 136, 138, 141

transect walks 89–90
transhumance *see* Niger
tree plantations 79–80, 98, 127
   see also nurseries; reforestation
tsunamis 93, 98
tube-wells 48

uncertainty 6–13, 18, 20, 24, 26, 31–2
United Nations (UN) 1, 4–5, 7
upazilla officers 44, 48, 50–1

variety seed selection 94–5
vegetable production 47, 63, 80
   see also farming
video 46, 84, 108
vulnerability 6, 13–18, 26–33
   Bangladesh 41
   Kenya 102, 105, 129
   Nepal 56
   Pakistan 82, 84
   Peru 149–50, 152, 154
   Sri Lanka 88, 89–90, 97

water management
   Kenya 104
   Nepal 61–2, 65
   Niger 118, 121–2
   Peru 151, 156, 158
   Sudan 131–45
water wells 51
watershed health 61–2
well construction 121–2, 129, 143
   see also tube-wells; water wells
winds 151
wood well construction 122
word of mouth awareness-raising 42
workshops 23
   Nepal 61, 65–6
   Niger 120
   Sri Lanka 95–6, 98

Yala monsoon climate 88, 95

WHSMITH EDUCATIONAL BOOKS

SCHONELL'S ESSENTIAL SPELLING LIST
workbook 3

FRED J SCHONELL

2 ESSENTIAL SPELLING

supply	avenue	government	enclose
support	diamond	madam	entry
attract	foundation	passage	magic
arrest	fuel	bandage	dye
shrink	cruelly	separate	ruin

☆ Learn the words, then use them to do the exercises.

☆ **OPPOSITES** Find words in the list which mean the opposite of these words.

1 repel _____   3 expand _____

2 free _____   7 exit _____

☆ **WORD HUNT** Which words in the list have these smaller words inside them?

1 tract _____   8   rate   _____

2 rink   _____   9   port   _____

3 ply   _____   10 and   _____

4 lose   _____   11 found _____

5 sage   _____   12 am   _____

6 over   _____   13 rest   _____

7 rue   _____   14 try   _____

☆ **MINI PUZZLE** Use words from the list to complete the puzzle.

Clues down
1 destroy
3 colouring substance
4 sorcery

Clues across
2 broad, tree-lined street
3 precious stone
5 source of heat or power

GROUP 5

latitude	employ	observe	violin
altitude	employer	observation	violet
minute	custom	desert	cricket
reduce	customer	slavery	clerk
refuge	fever	misery	onion

☆ Learn the words, then use them to do the exercises.

☆ **WORD PUZZLE** Use the word 'observation' to form eleven words from the list.

1. ___o___ hire
2. ___b___ notice
3. ___s___ client
4. ___e___ ancient team game
5. ___r___ great unhappiness
6. ___v___ detailed examination
7. ___a___ distance from the equator
8. ___t___ height
9. ___i___ a flower
10. ___o___ one who employs workers
11. ___n___ musical instrument

☆ **EXPLANATIONS** Complete these sentences with words from the list.

1 The word which means 'to make smaller' is _____.
2 The word which means 'habit' is _____.
3 A place of safety is a _____.
4 A _____ is a sixtieth part of an hour.
5 Another word for 'bondage' is _____.

☆ **WORD HUNT** Complete these sentences with words from the list.

1 The Danakil _____ is one of the hottest regions in the world.
2 When I cut the _____ my eyes watered.
3 I caught flu last month and had a high _____.
4 My sister has worked as a bank _____ for two years.

4 ESSENTIAL SPELLING

common	errand	arrange	neglect
collect	funnel	arranging	cruiser
connect	flannel	gallery	suspect
connection	channel	difficult	villain
command	current	umbrella	cannon

☆ Learn the words, then use them to do the exercises.

☆ **WORD PUZZLE** Use words from the list to complete the puzzle.

Clues across
1. order
3. smokestack
4. face cloth
6. wicked person
8. join
9. portable device for protection from rain
11. believe to be guilty
12. steady flow
14. pick up
16. course of a river
17. lack of attention
18. motor boat

Clues down
1. very large gun
2. hard
5. widespread
7. putting in order
8. link
10. organise
13. short trip to perform a task
15. covered balcony

GROUP 5

argue	citizen	surprise	complete
argument	century	purchase	estate
valuable	centre	purpose	minister
vegetable	central	further	receive
comfortable	hospital	scratch	deceive

☆ Learn the words, then use them to do the exercises.

☆ **WORD MATCH** Find words in the list which mean the same as these words.

1 shock _____
2 buy   _____
3 aim   _____
4 trick _____

5 get   _____
6 row   _____
7 full   _____
8 cosy   _____

☆ **WORD HUNT** Which words in the list have these smaller words inside them?

1 fur _____
2 rat _____
3 pit _____

4 get _____
5 ate _____
6 is   _____

☆ **EXPLANATIONS** Complete these sentences with words from the list.

1 If things are worth lots of money they are _____.
2 A middle position is a _____ one.
3 A _____ is an inhabitant of a town or city.
4 A period of a hundred years is a _____.
5 Another word for 'dispute' is _____.

☆ **WORD HUNT** Complete these sentences with words from the list.

1 The armchair was very _____ to sit in.
2 The people in the car crash were taken to the _____.
3 I was too tired to walk any _____.
4 The duke owns a large _____ in the country.
5 That jewel is very _____.

6 ESSENTIAL SPELLING

scare	insane	entrance	dodge
scarce	invade	performance	pledge
scarf	inspire	balance	divine
meanness	include	substance	ache
straight	introduce	lightning	headache

☆ Learn the words, then use them to do the exercises.

☆ **WORD PUZZLE**   Use the word 'headache' to help form eight words from the list.

1. brilliant light flash in a storm
2. bad pain in the head
3. worn round the neck
4. overrun
5. dramatic production
6. matter
7. continuous dull pain
8. god-like

☆ **OPPOSITES**   Find words in the list which mean the opposite of these words.

1 common _____
2 generosity _____
3 exit _____
4 exclude _____
5 crooked _____

☆ **WORD MATCH**   Find words in the list which mean the same as these words.

1 present _____
2 influence _____
3 evade _____
4 frighten _____
5 weigh _____
6 mad _____
7 promise _____

☆ **WORD HUNT**   Which words in the list have these smaller words inside them?

1 spire _____
2 lance _____
3 form _____
4 vine _____
5 ledge _____

GROUP 5

adopt	**guard**	**decide**	**mercy**
prompt	**guess**	**recite**	**multiply**
cupboard	**guinea**	**concert**	**ninth**
sponge	**tongue**	**fertile**	**fashion**
problem	**rogue**	**unite**	**thirsty**

☆ Learn the words, then use them to do the exercises.

☆ **EXPLANATIONS**   Complete this poem with words from the list.

1   A scoundrel is another name for _____.
2   If you are in the _____, you're in vogue.
3   If you are _____, then you'll need a drink.
4   If you've a _____, then you'll have to think.
5   If you're not sure, you'll have to make a _____.
6   If you've a wagging _____, you must talk less.
7   A _____ is a pound plus five pence more.
8   A _____ has some shelves behind a door.
9   If you are _____, you're not in the first eight.
10 If you are _____, then you will not be late.
11 A _____ land produces crops galore.
12 To _____ a sum will make it more.

☆ **MINI PUZZLE**   Use words from the list to complete the puzzle.

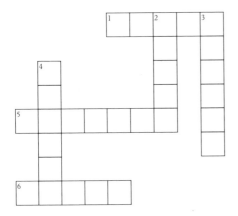

Clues down
  2 take another's child as one's own
  3 make a decision
  4 porous substance which soaks up water

Clues across
  1 shield
  5 performance of music
  6 pity

ESSENTIAL SPELLING

uniform	expense	ninety	period
perform	expensive	safety	roar
force	relative	surely	soar
skull	standard	entirely	cocoa
utmost	scholar	o'clock	situated

☆ Learn the words, then use them to do the exercises.

☆ **WORD MATCH** Find words in the list which mean the same as these words.

1 alike _____
2 dear _____
3 flag _____
4 placed _____
5 cost _____

6 wholly _____
7 kinsman _____
8 act _____
9 rise _____

☆ **WORD HUNT** Fill the gaps in the story with words from the list.

A King and his Cocoa

There was once a king who was so fond of (1) _____ that he would not eat or drink anything else. He drank so much that he fell ill. At twelve (2) _____ one night he woke up in great pain. He gave a loud (3) _____. "Bring me my doctor," he bellowed.

The doctor who was a great (4) _____ said, "Your majesty, you must (5) _____ yourself to stop drinking cocoa."
"Impossible," said the king.

"Then I cannot guarantee your (6) _____," said the doctor.
(7) "_____ you can get it into your (8) _____ that too much cocoa is bad for you. You must do your (9) _____ to give it up. If not, you will be dead within a week."

The king gave it up for a (10) _____ of (11) _____ days. After that he was well again. Now he only drinks cocoa at bed-time.

GROUP 5

general	embrace	threat	annual
generally	surface	weapon	cough
practical	furnace	forehead	ounce
natural	wasp	heaviness	soup
naturally	sleepiness	weariness	business

☆ Learn the words, then use them to do the exercises.

☆ **WORD PUZZLE** Use the word 'weariness' to help form nine words from the list.

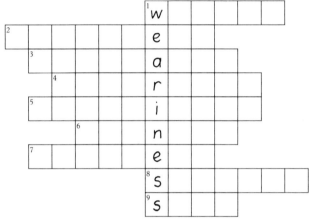

object used in fighting
tiredness
of course
usually
drowsiness
oven
trade or profession
top
liquid food made by boiling meat, etc.

☆ **EXPLANATIONS** Complete these sentences with words from the list.

1 A _____ is a winged, stinging insect.

2 Another word for _____ is 'widespread'.

3 A _____ shows intention to hurt or punish.

☆ **WORD PUZZLE** Use words from the list to complete the puzzle.

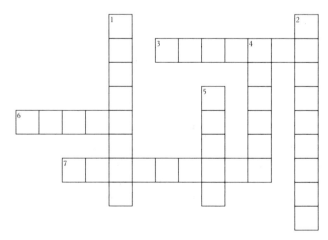

Clues down
1 brow
2 weight
4 yearly
5 one sixteenth of a pound

Clues across
3 clasp affectionately
6 expel air abruptly
7 adapted for use

ESSENTIAL SPELLING

bicycle	permission	orphan	sentence
biscuit	admission	geography	defence
juice	million	elephant	peace
statement	region	abundant	appeal
improvement	union	colony	instantly

☆ Learn the words, then use them to do the exercises.

☆ **WORD MAKER** Use words from the list to complete the words below.

1 _ _ _ ion   (area of land)
2 _ _ _ _ _ _ _ ion   (consent)
3 _ _ ion   (act of uniting into one)
4 _ _ _ _ _ ment   (an account of facts)
5 _ _ _ _ _ _ _ ment (act of making better)
6 _ _ _ _ _ ant   (plentiful)
7 _ _ _ _ _ ant   (animal with a trunk)

☆ **WORD PUZZLE** Use words from the list to complete the puzzle.

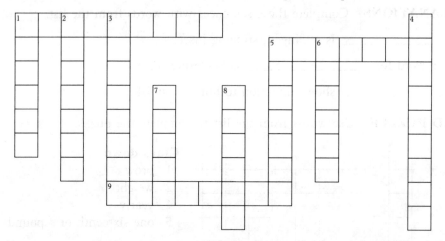

Clues down
1 ask for help
2 a thousand thousands
3 judgement
4 study of the earth's natural features
5 small, flat cake
6 territory occupied by settlers from another state
7 liquid obtained from plants or fruit
8 person whose parents are dead

Clues across
1 permission to enter
5 a two wheeled vehicle with pedals
9 biggest land animal

GROUP 5

author	August	brief	scenery
governor	autumn	priest	length
conductor	fault	shriek	depth
scent	pause	fierce	submit
scene	laundry	view	subtract

☆ Learn the words, then use them to do the exercises.

☆ **WORD PUZZLE** Use words from the list to complete the puzzle.

Clues down
2 ruler of a province
3 one who leads or guides
4 hesitate
5 take away
7 between summer and winter
9 distance downwards
10 defect

Clues across
1 eighth month of the year
6 washing
8 give in
11 short
12 writer
13 measurement from end to end

☆ **EXPLANATIONS** Complete these sentences with words from the list.

1 A _____ is a minister of the church.
2 Another word for 'perfume' is _____.
3 Another word for 'wild' is _____.
4 Another word for 'yell' is _____.

ESSENTIAL SPELLING

choice	curious	arouse	abrupt
rejoice	various	trousers	lungs
avoid	glorious	surround	fury
moisture	anxious	surrender	material
palm	worthy	wither	special

☆ Learn the words, then use them to do the exercises.

☆ **WORD MATCH**   Find words in the list which mean the same as these words.

1 shun _____
2 damp _____
3 curt _____
4 odd _____
5 tense _____

6 stir _____
7 trews _____
8 yield _____
9 cloth _____
10 diverse _____

☆ **WORD PUZZLE**   Use words from the list to complete the puzzle.

Clues down
1 deserving
2 encircle
3 of a particular kind

Clues across
4 renowned
5 feel great happiness
6 wild rage
7 tropical tree

☆ **WORD HUNT**   Complete these sentences with words from the list.

1 In Autumn the leaves _____ and fall from the trees.
2 I had a _____ of two cars but I couldn't make up my mind.
3 Doctors say that smoking is very bad for the _____.

GROUP 5   13

woollen	carriage	succeed	New Zealand
crooked	marriage	success	zone
loose	machine	successful	debt
foolish	acre	prayer	doubt
soldier	pearl	deny	

☆ Learn the words, then use them to do the exercises.

☆ **MINI PUZZLE** Use the word 'soldier' to form seven words from the list.

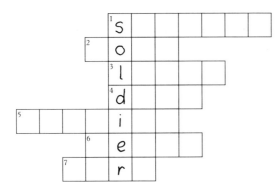

achievement

marked off area

not tight

declare as untrue

member of army

precious gem

measurement of land

☆ **WORD HUNT** Fill in the gaps in the story with words from the list.

Success Story

After the war a (1) _____ started a new life in (2) _____.
At first he was (3) _____ with his money and got into
(4) _____. Then an uncle left him some land on an industrial
(5) _____ in Wellington, the capital. On the land was a ramshackle building
called the (6) _____ House. In it the soldier installed a knitting
(7) _____. On this he produced (8) _____ goods. His friends
said that his business would not (9) _____ but the soldier had no
(10) _____ that (11) _____ would be his. The business
prospered. Now the soldier wished to marry. In answer to his (12) _____ he
met a young woman who liked him. They married and their (13) _____ was
just as (14) _____ as their business. The couple replaced the house with a
new factory. They bought a smart new home in the suburbs and every day the soldier drove
to work in a fine horsedrawn (15) _____. They named their new home after a
tumbledown building in Wellington. Can you guess what they called it?

14 ESSENTIAL SPELLING

personal	declare	portion	fund
liberal	decrease	proportion	minor
festival	decline	production	major
removal	determine	protection	majority
criminal	determination	introduction	traitor

☆ Learn the words, then use them to do the exercises.

☆ **WORD MATCH** Find words in the list which mean the same as these words.

1 fête _____
2 decide _____
3 private _____
4 part _____
5 crook _____
6 preface _____

7 defence _____
8 ratio _____
9 firmness _____
10 manufacturing _____
11 open-minded _____
12 announce _____

☆ **WORD PUZZLE** Use words from the list to complete the puzzle.

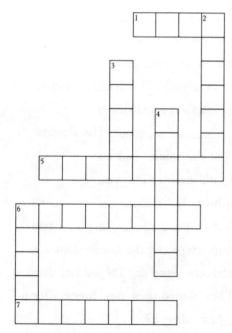

Clues down
2 refuse to accept
3 army officer
4 one who betrays his or her country
6 person under 18 years of age

Clues across
1 a reserve of money
5 reduction
6 the greater number
7 change of location

GROUP 6

greet	accurate	insult	keen
Greece	accuse	instruct	ghost
engineer	accustom	insert	skeleton
pioneer	announce	injure	cushion
career	addition	injury	income

☆ Learn the words, then use them to do the exercises.

☆ **WORD HUNT** Which words in the list have these smaller words inside them?

1 ounce _____
2 come _____
3 gin _____
4 rate _____
5 custom _____
6 it _____
7 use _____
8 one _____
9 let _____
10 are _____
11 jury _____
12 host _____

☆ **EXPLANATIONS** Complete these sentences with words from the list.

1 Another word for 'hurt' is _____.
2 A _____ is a kind of pillow.
3 When you _____ people you teach them.
4 'To put something in' means to _____ it.
5 An 'offensive remark' is an _____.

☆ **WORD MAKER** Use words from the list to complete the words below.

1 _ ee _   (enthusiastic)
2 _ _ ee _ _   (a country in Europe)
3 _ _ ee _   (to meet or welcome)
4 _ _ _ _ ee _   (someone who opens the way)
5 acc _ _ _ _ _ _ (correct)
6 acc _ _ _ _   (to blame)
7 acc _ _ _ _ _ _ (get used to)

16 ESSENTIAL SPELLING

contract	prefer	wholesome	convince
control	preferred	enterprise	poison
consent	conferred	therefore	coil
contempt	grudge	wireless	wisdom
conclude	lodging	grateful	condemn

☆ Learn the words, then use them to do the exercises.

☆ **WORD HUNT** Which words in the list have these smaller words inside them?

1 sent _____
2 referred _____
3 less _____
4 act _____
5 rise _____
6 in _____
7 hole _____
8 son _____
9 tempt _____
10 ate _____
11 here _____
12 oil _____

☆ **EXPLANATIONS** Complete these sentences with words from the list.

1 When you _____ a thing you end it.
2 'Curb' is another word for _____.
3 _____ means to like something better.
4 If people _____ about something, they consulted together.
5 Those who have _____ act with common sense and understanding.
6 To _____ is to express strong disapproval.

☆ **MINI PUZZLE** Use the word 'condemn' to form seven words from the list.

1 c — twist into spiral shape
2 o — dwelling
3 n — bring to an end
4 d — resentment
5 e — to like better
6 m — act of despising
7 n — to restrain

GROUP 6

hesitate	continent	stage	debate
delicate	fragment	garage	student
candidate	regiment	average	confident
certificate	experiment	discourage	camera
navigate	cement	baggage	remedy

☆ Learn the words, then use them to do the exercises.

☆ **WORD MATCH** Find words in the list which mean the same as these words.

1 fragile _____

2 piece _____

3 prevent _____

4 pause _____

5 test _____

6 sail _____

7 cure _____

8 sure _____

☆ **WORD HUNT** Complete these sentences with words from the list.

1 Asia is the world's largest _____.

2 On passing his examination the _____ was awarded a _____.

3 The _____ _____ allowance on a plane is 44lbs.

4 The prime minister will open today's _____ in Parliament.

5 The new house has a two-car _____ with up-and-over doors.

6 The Labour _____ asked me to vote for her at the General Election.

7 To make concrete you mix _____ with water, sand and gravel.

8 I am _____ that with my new _____ I will get some excellent photographs.

9 He is a fine soldier and the _____ is proud of him.

10 The lights went out on _____ during the last act of the play.

ESSENTIAL SPELLING

treaty	lecture	nervous	recent
treatment	agriculture	prosperous	recently
ornament	temperate	tremendous	volcano
instrument	temperature	ridiculous	couch
prominent	puncture	jealous	route

☆ Learn the words, then use them to do the exercises.

☆ **WORD HUNT**   Complete these sentences with words from the list.

1 The _____ patient had a _____ fear of spiders and leapt from his _____ at the sight of one.

2 The _____ merchant gave _____ sums of money to charity.

3 Jones, a born failure, was very _____ of the success of his younger sister.

4 The Minister of _____, himself a farmer, is well qualified to _____ on the subject.

5 In _____ countries the _____ is neither very hot nor very cold.

6 The cyclist had to repair a front tyre _____ and so lost a lot of time in the race.

7 A _____ doctor claims that the best _____ for most diseases is a healthy diet.

8 The telephone is not just a useful _____; it also serves as an _____ in many homes.

9 The _____, which erupted _____, has caused no loss of life or property yet.

☆ **WORD HUNT**   Which words in the list have these smaller words inside them?

1 eat   _____   3 out   _____
2 cent _____   4 mend _____

☆ **WORD MATCH**   Find words in the list which mean the same as these words.

1 rich   _____   3 laughable _____
2 marvellous _____

GROUP 6   19

sensible	ignorant	ceiling	hatred
responsible	ignorance	perceive	sacred
visible	abundance	deceit	witch
invisible	attendance	deceitful	wretched
rifle	appearance	earthquake	wrinkle

☆ Learn the words, then use them to do the exercises.

☆ **WORD HUNT**   Which words in the list have these smaller words inside them?

1 pear _____   5 if _____

2 itch _____   6 ink _____

3 on _____   7 ant _____

4 etch _____   8 ten _____

☆ **WORD PUZZLE**   Use the word 'invisible' to help form nine words from the list.

fraud

lack of knowledge

evident

fraudulent

not able to be seen

notice

showing good sense

upper limit of a room

vibrations at the earth's surface

☆ **WORD MAKER**   Use words from the list to complete the words below.

1 _ _ _ red   (intense dislike)

2 _ _ _ red   (holy)

3 _ _ _ _ _ _ance (in plenty)

4 _ _ _ _ _ _ _ance (outward show)

5 _ _ _ _ _ _ance (lack of knowledge)

ESSENTIAL SPELLING

telescope	national	conversation	phrase
telegram	cathedral	consideration	fatal
telegraph	principal	sensation	section
photograph	punctual	combination	intention
physical	continually	ventilation	choir

☆ Learn the words, then use them to do the exercises.

☆ **WORD HUNT**   Complete these sentences with words from the list.

1 The _____ cause of _____ road accidents is exceeding the speed limit.

2 Canterbury _____ is one of our great _____ monuments.

3 The patient complained _____ about a constant _____ of giddiness.

4 In deep coal mines _____ is a very important _____.

5 I have never been late for school and it is my _____ to be equally _____ when I start work.

6 The _____ of lack of food and poor living conditions turned the prisoner of war into a _____ wreck.

7 Nelson is said to have clapped the _____ to his blind eye and said, "I see no ships".

8 During the storm a _____ pole blew down, blocked the road and caused a huge traffic jam.

9 I was delighted to receive a _____ informing me that I had won the pools.

10 That old _____ of me is not a very good likeness.

☆ **MINI PUZZLE**   Use the word 'choir' to help form five words from the list.

1. c _____ talk
2. h _____ small group of words
3. o _____ group of singers
4. i _____ part
5. r _____ most important

GROUP 6

domestic	tenant	fragrant	item
athletic	vacant	insurance	ideal
heroic	tyrant	assistance	pilot
majestic	elegant	remembrance	pistol
tropics	extravagant	circumstance	seize

☆ Learn the words, then use them to do the exercises.

☆ **WORD PUZZLE** Use words from the list to complete the puzzle.

Clues down
1 empty
2 muscular
4 memento
6 homely
9 financial protection for health, property, etc.
10 grasp
11 one who operates an aircraft
13 oppressive ruler
15 perfect

Clues across
3 sweet smelling
5 tasteful in dress
7 area round the equator
8 article
12 condition
13 one who pays rent
14 dignified
16 help
17 spending money excessively

22 ESSENTIAL SPELLING

miserable	despair	convert	liable
reasonable	despise	concern	reliable
capable	description	convey	angle
probable	destruction	witness	trifle
probably	energy	cleanliness	muscle

☆ Learn the words, then use them to do the exercises.

☆ **WORD HUNT** Complete these sentences with words from the list.

1 My _____ of the attacker is _____ not _____ as I only caught a brief glimpse of him.

2 It is _____ that the damaged _____ will recover if you take a _____ amount of exercise.

3 You can imagine how _____ life was after the _____ of the city by the earthquake.

4 Please _____ to the travel agent our thanks for her most _____ arrangements for our holiday.

5 An old proverb says that '_____ is next to godliness'.

6 The duty of a _____ is to tell the truth, the whole truth and nothing but the truth.

7 Sentencing the traitor, the judge said, "Your fellow citizens will always _____ you for your treachery."

8 The main _____ of alchemists of old was to _____ base metals into gold.

9 He had very little _____ after his illness so the doctor gave him a tonic.

10 The man was declared _____ by the court and he had to pay the damages.

☆ **WORD HUNT** Which words in the list have these smaller words inside them?

1 pair _____   3 an _____
2 rifle _____

influence	famine	reception	absence
presence	medicine	ambition	pretence
evidence	genuine	satisfaction	umpire
residence	granite	objection	crime
reference	definite	instruction	circus

☆ Learn the words, then use them to do the exercises.

☆ **WORD HUNT** Fill in the gaps in the two letters with words from the list.

Dear Sir,

All (1) _genuine_ animal lovers consider it a (2) _crime_ for animals to be employed in the (3) _circus_. We have the strongest (4) _objection_ to the (5) _instruction_ given to circus animals who get no (6) _satisfaction_ from their training. It is our (7) _ambition_ to stamp out this exploitation. We call on all animal lovers to give a hostile (8) _reception_ to circuses whenever they come to town.

Yours faithfully
A Clark (President: Animal Lovers Guild)

Dear John,

With (9) _reference_ to your letter of last week, I am delighted that you are going to take up (10) _residence_ in Africa. Your (11) _presence_ there can only (12) _influence_ our work for the better. I make no (13) _pretence_ that the (14) _absence_ of a resident director has been a (15) _definite_ handicap in the past. There is (16) _evidence_ that we are getting the (17) _famine_ under control though (18) _medicine_ for the sick is in short supply.

Yours sincerely
F C Bates (Controller: African Relief Agency)

24 ESSENTIAL SPELLING

salute	detail	acquire	luxury
distribute	deposit	acquaint	burden
gratitude	develop	acquainted	swollen
destitute	strength	acquaintance	execute
volume	strengthen	disappoint	abbey

☆ Learn the words, then use them to do the exercises.

☆ **WORD MATCH** Find words in the list which mean the same as these words.

1 needy _____   7 grow _____
2 greet _____   8 thanks _____
3 allot _____   9 friend _____
4 put down _____   10 get _____
5 size _____   11 item _____
6 kill _____   12 load _____

☆ **WORD HUNT** Fill in the gaps in this letter with words from the list.

The Editor: Abbeytown Gazette

Dear Sir,

May I use your columns to (1) _____ your readers with some of the problems facing those trying to restore our local (2) _____ church? During recent storms the (3) _____ river flooded the site. We appeal for young volunteers with the (4) _____ to help in the work of digging drainage channels. This is essential before we can begin to (5) _____ the foundations themselves. The work is no (6) _____, as anyone (7) _____ with working ankle deep in mud will be aware. I hope you won't (8) _____ us.

Yours faithfully

Rev. Canon W Williams

☆ **WORD HUNT** Which words in the list have these smaller words inside them?

1 sit _____   3 den _____
2 tribute _____   4 cute _____

GROUP 6   25

pattern	release	offend	response
messenger	reveal	oppose	resign
traveller	reflect	opposite	design
challenge	reserve	opposition	oppress
college	remainder	application	approve

☆ Learn the words, then use them to do the exercises.

☆ **WORD HUNT**   Complete these sentences with words from the list.

1 I sent off my _____ a week ago but have had no _____ so far.
2 The _____ of _____ is 'disapprove'.
3 Over half the prisoners obtained their _____ last week and the _____ expect to be freed next week.
4 The writings of Stevenson, who was a great _____ in France, _____ his love and knowledge of the land and its people.
5 "I warn you," said the judge, "that if you _____ once more, you will go to prison for a very long time."
6 My mother is so clever at dressmaking that she never uses a _____ of any kind. She always creates her own _____.
7 The _____ has entered a team to appear in the TV show 'University _____'.
8 The letter was urgent so I sent a _____ by motorbike to deliver it.

☆ **WORD MAKER**   Use words from the list to complete the words below.

1 opp _ _ _ _ _   (treat unjustly)   4 re _ _ _ _ _   (give up a position)
2 opp _ _ _   (resist)   5 re _ _ _ _ _ _   (to book)
3 opp _ _ _ _ _ _ _ _ (resistance)   6 re _ _ _ _   (to show)

☆ **WORD HUNT**   Which words in the list have these smaller words inside them?

1 all _____   3 main _____
2 ease _____   4 end   _____

ESSENTIAL SPELLING

flavour	generous	innocent	practise
vapour	numerous	independent	impudent
rumour	enormous	excitement	yield
occasion	mischievous	advertise	shield
occasionally	marvellous	advertisement	pierce

☆ Learn the words, then use them to do the exercises.

☆ **WORD HUNT** Complete these sentences with words from the list.

1 Jones is very _____ and will not accept help from anybody.

2 I know that _____ Mary is _____ but she is really quite _____ and means no harm.

3 Billy is very polite. Indeed I have known him to be _____ on only one _____.

4 We had _____ replies to our _____ which really does prove that it pays to _____.

5 She gave a _____ display of ice skating which kept the spectators in a fever of _____ from first to last.

☆ **MINI PUZZLE** Use the word 'shield' to help form five words from the list.

```
1 _ _ _ _ s _ _        work at repeatedly
      2   h              to protect
3 _ _ _ _ _ _ _          
        4 e _ _          to make public
          5 _ l _        make a hole in
            d            to give way
```

work at repeatedly
to protect
to make public
make a hole in
to give way

☆ **WORD MAKER** Use words from the list to complete the words below.

1 _ _ _ ou _   (gaseous substance)
2 _ _ _ ou _   (hearsay)
3 _ _ _ _ ou _   (taste)
4 _ _ _ _ _ ou _   (huge)
5 _ _ _ _ _ ou _   (not mean)

GROUP 6

invalid	digest	provoke	frigid
invention	digestion	proclaim	civil
patient	soul	pronounce	civilised
impatient	mould	proceed	operation
impatience	poultry	opinion	heir

☆ Learn the words, then use them to do the exercises.

☆ **WORD PUZZLE**   Use the word 'impatience' to help form ten words from the list.

```
1 _ _ _ i _ _ _ _ _       untrue
  2 _ m _ _ _ _ _ _        lacking patience
  3   p _ _ _ _ _          point of view
  4   a _ _ _ _            irritation
 5 _ _ t _ _ _ _           surgery to repair damage
6 _ _ i _ _ _ _ _          creation of new ideas
    7 _ e _ _ _ _          breaking up of food in body
    8 _ n _ _ _ _          say words the right way
    9 _ c _ _ _ _ _        announce publicly
10 _ _ _ _ e _             cultured
```

☆ **WORD HUNT**   Complete these sentences with words from the list.

1 The polar regions are the earth's _____ zones.
2 The _____ was too weak to _____ his food.
3 We ran out of petrol miles from anywhere and could _____ no further that night.
4 We did not want to _____ the police when they questioned us so we were very _____ in our replies.
5 We get our eggs from a nearby _____ farm.
6 I threw the bread out because it had _____ on it.

☆ **WORD MAKER**   Use words from the list to complete the words below.

1 pro _ _ _ _ _   (to stir to action)   3 pro _ _ _ _ _ _   (to announce)
2 pro _ _ _ _ _   (to continue)

ESSENTIAL SPELLING

commit	comrade	pension	solve
commence	complaint	provision	dissolve
recommend	humour	decision	wholly
recollect	endeavour	conclusion	annoy
shipping	tobacco	division	annoyed

☆ Learn the words, then use them to do the exercises.

☆ **WORD HUNT**   Which words in the list have these smaller words inside them?

1 pen   _____   7   pin   _____
2 collect _____   8   lain   _____
3 vision   _____   9   so   _____
4 solve   _____   10   annoy   _____
5 mend   _____   11   our   _____
6 holly   _____

☆ **WORD HUNT**   Fill in the gaps in this letter with words from the list.

Dear _____,

Your committee has made the (2) _____ to ban (3) _____ smokers from this club. The ban will (4) _____ on the 1st of next month. We reached this (5) _____ after taking the views of all members into account. It was made clear that the few who smoke pollute the atmosphere and gravely (6) _____ the majority of other members. Smoking will therefore be an offence against club rules. Those who (7) _____ it will have their membership revoked.

Yours sincerely

J Brown (Chairman: Oldtown British Legion Club)

☆ **WORD MATCH**   Find words in the list which mean the same as these words.

1 end   _____
2 completely _____
3 friend   _____

GROUP 6   29

poverty	margin	transform	arrival
mutiny	origin	translate	Spain
variety	original	character	dainty
society	moral	programme	quaint
sacrifice	crystal	sandwich	quench

☆ Learn the words, then use them to do the exercises.

☆ **OPPOSITES**   Find words in the list which mean the opposite of these words.

1 departure _____   4 riches _____
2 commonplace _____   5 sinful _____
3 sameness _____

☆ **WORD PUZZLE**   Use words from the list to complete the puzzle.

Clues down
2 delicate
3 written list of events
4 beginning
5 express in another language
6 offering to a god
9 group organised to share a common interest
10 put out

Clues across
1 two slices of bread with a filling between
7 change completely
8 brilliant type of glass
11 old fashioned
12 rebellion against superior officers
13 a person's nature

ESSENTIAL SPELLING

create	suitable	Chinese	scheme
creation	creditable	interfere	pursue
emigrate	honourable	supreme	pursuit
emigrant	peaceable	extreme	mourn
obstinate	manageable	extremely	source

☆ Learn the words, then use them to do the exercises.

☆ **WORD MATCH** Find words in the list which mean the same as these words.

1 stubborn _____

2 meddle _____

3 chase _____

4 worthy _____

5 calm _____

6 plan _____

7 fit _____

8 produce _____

☆ **WORD HUNT** Fill the gaps in the passage with words from the list.

Hong Kong

Many mainland (1) _____ (2) _____ illegally to Hong Kong.

Though housing there is (3) _____ limited, it is at least

(4) _____ compared with conditions in China itself.

The (5) _____ will probably start her new life in (6) _____

poverty. She will hope to work hard and save enough to start her own business.

☆ **MINI PUZZLE** Use the word 'create' to help form six words from the list.

1. c — origin
2. r — grieve
3. e — of highest quality
4. a — act of bringing into existence
5. t — appropriate
6. e — plan

GROUP 6

intelligent	judgement	thorough	torrent
intelligence	parliament	sustain	squirrel
difference	incident	maintain	essay
offence	magnificent	portrait	type
apparent	compliment	quarry	typewriter

☆ Learn the words, then use them to do the exercises.

☆ **WORD MATCH**   Find words in the list which mean the same as these words.

1 event _____   4 prey _____

2 verdict _____   5 clear _____

3 crime _____   6 keep _____

☆ **WORD HUNT**   Fill in the gaps in this account with words from the list.

Super Squirrel

You may think that squirrels do not have much (1) _____. Well, think again! I can tell you of a (2) _____ who was so (3) _____ that he won a national competition for (4) _____ writing. One (5) _____ between this squirrel and his fellows was that this one could (6) _____. Not only could he operate a (7) _____ with ease, he was also a gifted artist. He once won a competition for (8) _____ painting. One critic described his painting as (9) _____ — not a bad (10) _____, that, for a squirrel!

☆ **MINI PUZZLE**   Use words from the list to complete the puzzle.

Clues down
1 support
2 a very fast stream
3 carried out with care

Clues across
4 a body which makes a country's laws

32 ESSENTIAL SPELLING

vulgar	boundary	convenient	existence
similar	tributary	convenience	consequence
irregular	missionary	experience	psalm
circular	salary	obedient	attach
military	extraordinary	obedience	fowl

☆ Learn the words, then use them to do the exercises.

☆ **WORD MATCH** Find words in the list which mean the same as these words.

1 uneven _____   7 round _____

2 alike _____   8 handy _____

3 unusual _____   9 border _____

4 rude _____   10 knowledge _____

5 pay _____   11 being _____

6 join _____   12 result _____

☆ **WORD HUNT** Fill the gaps in the passage with words from the list.

The Missionary

The (1) _____ set up his mission on a (2) _____ of the Amazon River. The site was on the state (3) _____ far from any town. He would have preferred the greater (4) _____ of working in a town or village, where roads and telephones would make his (5) _____ more bearable. However, he had taken a vow of (6) _____ to his superiors. They wanted the mission in a certain place and no other. Because he was (7) _____, and for no other reason, he went where he was told.

☆ **MINI PUZZLE** Use the word 'psalm' to help form three words from the list.

a song in the bible
of or for soldiers
a bird like a hen

GROUP 6

ascend	plague	obstacle	rear
descend	league	miracle	breathe
science	fatigue	spectacle	cease
scissors	disguise	violent	conceal
intimate	disaster	permanent	awkward

☆ Learn the words, then use them to do the exercises.

☆ **WORD HUNT**   Fill the gaps in the passage with words from the list.

Football Violence

We cannot (1) _____ the fact that (2) _____ football is in a mess. Bigger crowds are needed but the (3) _____ of (4) _____ behaviour on the terraces has caused millions of fans to (5) _____ attending. When this (6) _____ has been overcome, all lovers of the game will (7) _____ a great sigh of relief.

☆ **WORD PUZZLE**   Use the word 'permanent' to help form nine words from the list.

1. widespread contagious disease
2. go down
3. instrument for cutting cloth, etc
4. intended to last for a long time
5. clumsy
6. private, secret
7. a special branch of knowledge
8. hide
9. tiredness

☆ **EXPLANATIONS**   Complete these sentences with words from the list.

1 A _____ is a supernatural event.
2 'To go up' means to _____.
3 Another word for 'breed' is _____.
4 To _____ something means 'to hide it'.

ESSENTIAL SPELLING

university	social	precious	charity
opportunity	artificial	gracious	ability
possibility	especially	delicious	brooch
responsibility	musician	suspicious	stomach
curiosity	triumph	suspicion	wreath

☆ Learn the words, then use them to do the exercises.

☆ **BOYS' NAMES** Write words from the list in which the names of these boys appear.

1 Al _____
2 Tom _____
3 Ian _____

☆ **WORD MATCH** Find words in the list which mean the same as these words.

1 costly   _____
2 kind   _____
3 wary   _____
4 chance   _____
5 tasty   _____
6 skill   _____
7 false   _____
8 bounty   _____
9 likelihood _____
10 blame   _____

☆ **WORD HUNT** Fill the gaps in the poem with words from the list.

The Lying in State

His body was laid out beneath

A violin shaped holly (1) _____.

His final (2) _____ came today

When thousands went out of their way

To pay their last respects to one

Whom fame had brightly shone upon.

(3) _____ they came to see

Just out of (4) _____,

The (5) _____ which he'd worn constantly.

Together with his (6) _____ ring

The gifts of an admiring king

Who'd never had the the least (7) _____

That he was not a great (8) _____.

GROUP 6

religion	necessary	exceed	solemn
religious	necessity	exclude	prophet
previous	furious	exception	scarlet
victorious	serious	expedition	excursion
industrious	behaviour	explode	saviour

☆ Learn the words, then use them to do the exercises.

☆ **VOWEL SPOTTER**   Write words from the list which contain five vowels.

1 _____   4 _____

2 _____   5 _____

3 _____

☆ **WORD MATCH**   Find words in the list which mean the same as these words.

1 grave   _____   6 former   _____

2 omit   _____   7 raging   _____

3 outing   _____   8 needed   _____

4 need   _____   9 red   _____

5 conduct   _____   10 devout   _____

☆ **EXPLANATIONS**   Complete these sentences with words from the list.

1 The word which means 'belief in' and 'worship of God' is _____.

2 One who foretells the future is a _____.

3 'To burst with great force' is to _____.

4 To take _____ to a thing is 'to object to it'.

5 A word for 'to go beyond' is to _____.

6 Another word for 'grave' is _____.

7 A _____ is a person who saves another.

☆ **WORD MAKER**   Use words from the list to complete the words below.

1 ex _ _ _ _   (to go beyond)

2 ex _ _ _ _ _ _ _ (not according to a rule)

3 ex _ _ _ _ _   (to burst)

4 ex _ _ _ _ _ _ _ _ (trip made for a special purpose)

ESSENTIAL SPELLING

distinguish	**system**	**persevere**	**abolish**
extinguish	**sympathy**	**atmosphere**	**leisure**
persuade	**mystery**	**electricity**	**sovereign**
establish	**delivery**	**interrupt**	**urge**
diminish	**discovery**	**settler**	**urgent**

☆ Learn the words, then use them to do the exercises.

☆ **WORD HUNT** Which words in the list have these smaller words inside them?

1 stem _____   7 reign _____

2 livery _____   8 sphere _____

3 stab _____   9 cover _____

4 city _____   10 sure _____

5 path _____   11 severe _____

6 sting _____

☆ **WORD PUZZLE** Use the word 'extinguish' to help form ten words from the list.

colonist
put out
unexplained event
stamp out
needing speedy attention
insist on
break in upon
a king or queen
cause to believe
lessen

☆ **WORD MAKER** Use words from the list to complete the words below.

1 _ _ _ _ _ _ _ish   (to put out)

2 _ _ _ _ _ ish   (to become smaller)

3 _ _ _ _ _ _ ish   (to set up)

4 _ _ _ _ _ _ _ _ish   (to see or hear plainly)

GROUP 6

pressure	affection	scribble	appoint
assure	affectionate	ruffle	affair
assume	attraction	rubbish	twilight
assent	accomplish	summit	coarse
assemble	accompany	traffic	hoarse

☆ Learn the words, then use them to do the exercises.

☆ **WORD HUNT**   Which words in the list have these smaller words inside them?

1 action _____   7 ate   _____
2 ruff   _____   8 sure   _____
3 press   _____   9 air   _____
4 rib   _____   10 rub   _____
5 company _____   11 point   _____
6 light   _____   12 oar   _____

☆ **MINI PUZZLE**   Use words from the list to complete the puzzle.

Clues down
1 achieve
4 vehicles coming and going

Clues across
2 fondness
3 highest point
5 rough

☆ **WORD MAKER**   Use words from the list to complete the words below.

1 ass _ _ _   (take for granted)   3 ass _ _ _ (agreement)
2 ass _ _ _ _ _   (come together)   4 ass _ _ _ (convince)

ESSENTIAL SPELLING

nursery	organize	occupation	collection
jewellery	organization	congregation	attractive
machinery	realize	preparation	novel
prospect	recognize	separation	marvel
satisfy	horizon	champion	cancel

☆ Learn the words, then use them to do the exercises.

☆ **WORD HUNT** Complete these sentences with words from the list.

1 _____ school teaching is a very demanding _____.

2 As we crossed the channel one could at last _____ the White Cliffs of Dover on the _____.

3 The _____ for the retiring vicar was a record even for a _____ as generous as ours.

4 The _____ declared, "I owe my victory to hard training and careful _____."

5 "I _____," said the author, "that I cannot please all my readers but I hope my new _____ will _____ most of them."

6 I finally turned down the job in Australia because I could not face the _____ of _____ from my family.

7 If you decide to join our _____ we can offer you the most _____ working conditions.

8 I asked the newsagent to _____ my papers as I was going on holiday for two weeks.

9 Next year the industrial museum is going to _____ an exhibition of textile _____.

10 Imitation _____ is a _____ to me — you can't tell a lot of it from real gems.

☆ **WORD HUNT** Complete these sentences with words from the list.

1 The word with the most vowels is _____.

2 The word with the most consonants is _____.

GROUP 6

celebrate	passion	interior	imitate
celebration	impression	exterior	imitation
illustrate	discussion	inferior	soothe
chocolate	possession	superior	indulge
immediately	correspond	senior	indulging

☆ Learn the words, then use them to do the exercises.

☆ **WORD HUNT**   Complete these sentences with words from the list.

1 If you go on _____ your _____ for _____ you will grow very fat indeed.
2 It is amazing how quickly a dummy can _____ a crying baby.
3 When the house came into his _____ he _____ asked all his friends round for a _____.
4 A famous artist was engaged to _____ the book.
5 I didn't think anyone could _____ my signature but when I saw the forgery I realised what a clever _____ it was.

☆ **WORD MATCH**   Find words in the list which mean the same as these words.

1 write   _____   5 outside   _____
2 inside   _____   6 higher   _____
3 lower   _____   7 older   _____
4 debate   _____   8 gratify   _____

☆ **WORD HUNT**   Which words in the list have these smaller words inside them?

1 press   _____   3 respond   _____
2 mediate   _____   4 rate   _____

☆ **WORD MAKER**   Use words from the list to complete the words below.

1 _ _ _ ior   (higher in rank)
2 _ _ _ _ _ ior   (the inside)
3 _ _ _ _ _ ior   (poor in quality)

ESSENTIAL SPELLING

benefit	audience	exhaust	launch
benefited	authority	exhibit	breadth
profited	clause	exhibition	suburb
partner	applaud	register	council
privilege	cautious	nourish	album

☆ Learn the words, then use them to do the exercises.

☆ **WORD PUZZLE** Use words from the list to complete the puzzle.

Clues across
2 cheer
5 district on outskirts of a town
7 hearers
8 feed
9 to show
10 profited
11 list of names
12 start off
13 use up
15 advantage
16 body which governs a town

Clues down
1 companion
3 a right
4 careful
6 gained
7 a book for stamps, photos, etc.
14 power to control others
17 a display
18 part of a document (eg a will)
19 width

GROUP 6   41

nonsense	knowledge	instructor	dense
suspense	acknowledge	conquer	haughty
condense	postpone	conqueror	slaughter
immense	envelope	radiator	curve
intense	liquid	mayor	disease

☆ Learn the words, then use them to do the exercises.

☆ **WORD HUNT**   Fill the gaps in the passage with words from the list.

I wrote to my (1) _____ asking him to (2) _____ my driving test. I explained that my car (3) _____ was leaking and thought that the loss of (4) _____ might be a danger. It took him a week to (5) _____ my letter. He said it was (6) _____ to cancel a test for such a reason.

☆ **WORD PUZZLE**   Use words from the list to complete the puzzle.

Clues across
3 savage killing
6 of great force
7 reduce in size
8 unhealthy condition
9 victor

Clues down
1 arrogant
2 bend
3 tension
4 wrapper
5 defeat

☆ **WORD HUNT**   Fill the gaps in the limerick with words from the list.

My Friend

My friend is amazingly (1) _____.

His ignorance is quite (2) _____.

The (3) _____ he's got

Wouldn't fill a pint pot.

He hasn't a ha'p'orth of sense.

ESSENTIAL SPELLING

peculiar	associate	scorch	ancient
familiar	association	horror	hymn
brilliant	appreciation	stubborn	column
Spaniard	official	Christian	bruise
Spanish	sufficient	positive	survive

☆ Learn the words, then use them to do the exercises.

☆ **WORD MATCH** Find words in the list which mean the same as these words.

1 club   _____
2 line   _____
3 vivid   _____
4 certain   _____
5 odd   _____

6 mix   _____
7 thanks   _____
8 enough   _____
9 old   _____
10 burn   _____

☆ **WORD HUNT** Fill the gaps in the passage with words from the list.

The Christians in Rome

In (1) _____ Rome, Christians had to be very (2) _____ indeed to (3) _____ the early persecutions. We read with (4) _____ of the cruelty they suffered at the hands of the tyrant, Nero. Later, however, the (5) _____ faith became the (6) _____ religion of the empire.

☆ **WORD HUNT** Fill the gaps in the limerick with words from the list.

Sing Song

A (1) _____, whom we nicknamed Jim,

Played a not too (2) _____ (3) _____.

To his (4) _____ guitar

We sang 'Tra la la la',

Then we dived in the pool for a swim.

courteous	rescue	skilful	siege
courageous	virtue	pitiful	salmon
museum	issue	welfare	Egypt
unusual	procession	fulfil	absurd
suggest	succession	ceremony	theatre

☆ Learn the words, then use them to do the exercises.

☆ **WORD HUNT** Complete the three passages with words from the list.

At the (1) _____ of Alexandria in (2) _____ the plight of the

people was (3) _____. (4) _____ to the last, they defied a

(5) _____ of fierce attacks without regard to their own

(6) _____.

The mayor led a civic (1) _____ for the opening (2) _____.

He was surprisingly (3) _____ to those who booed his speech. He politely

reminded them that he was there to (4) _____ his election pledge to rebuild

the town's (5) _____ and open-air (6) _____.

A Fishy Tale

That most (1) _____ fish, the (2) _____,

Is very (3) _____ at backgammon.

And, though I know it sounds (4) _____,

He's never known to break his word.

Of every (5) _____ he's possessed.

In fact, I venture to (6) _____

He is the finest of all fish.

(He also makes a tasty dish).

☆ **WORD MAKER** Use words from the list to complete the words below.

1 _ _ _ ue   (matter to be discussed)   5 _ _ _ _ ful   (deserving sympathy)

2 _ _ _ _ _ ue   (to free from danger)   6 _ _ _ _ _ _ eous   (polite)

3 _ _ _ _ _ ue   (quality of goodness)   7 _ _ _ _ _ _ _ eous (brave)

4 _ _ _ _ ful   (able to do something well)

Answers

2 OPPOSITES (1) attract (2) arrest (3) shrink
(4) entry
WORD HUNT (1) attract (2) shrink (3) supply
(4) enclose (5) passage (6) government
(7) cruelly (8) separate (9) support (10) bandage
(11) foundation (12) diamond (13) arrest
(14) entry
MINI PUZZLE **Clues down** (1) ruin (3) dye
(4) magic
Clues across (2) avenue (3) diamond (5) fuel

3 WORD PUZZLE (1) employ (2) observe
(3) customer (4) cricket (5) misery
(6) observation (7) latitude (8) altitude
(9) violet (10) employer (11) violin
EXPLANATIONS (1) reduce (2) custom
(3) refuge (4) minute (5) slavery
WORD HUNT (1) Desert (2) onion (3) fever
(4) clerk

4 WORD PUZZLE **Clues down** (1) cannon
(2) difficult (5) common (7) arranging
(8) connection (10) arrange (13) errand
(15) gallery
Clues across (1) command (3) funnel
(4) flannel (6) villain (8) connect (9) umbrella
(11) suspect (12) current (14) collect
(16) channel (17) neglect (18) cruiser

5 WORD MATCH (1) surprise (2) purchase
(3) purpose (4) deceive (5) receive (6) argument
(7) complete (8) comfortable
WORD HUNT (1) further (2) scratch
(3) hospital (4) vegetable (5) estate
(6) surprise, minister
EXPLANATIONS (1) valuable (2) central
(3) citizen (4) century (5) argument
WORD HUNT (1) comfortable (2) hospital
(3) further (4) estate (5) valuable

6 WORD PUZZLE (1) lightning (2) headache
(3) scarf (4) invade (5) performance
(6) substance (7) ache (8) divine
OPPOSITES (1) scarce (2) meanness
(3) entrance (4) include (5) straight
WORD MATCH (1) introduce (2) inspire
(3) dodge (4) scare (5) balance (6) insane
(7) pledge
WORD HUNT (1) inspire (2) balance
(3) performance (4) divine (5) pledge

7 EXPLANATIONS (1) rogue (2) fashion
(3) thirsty (4) problem (5) guess (6) tongue
(7) guinea (8) cupboard (9) ninth (10) prompt
(11) fertile (12) multiply
MINI PUZZLE **Clues down** (2) adopt
(2) decide (4) sponge
Clues across (1) guard (5) concert (6) mercy

8 WORD MATCH (1) uniform (2) expensive
(3) standard (4) situated (5) expense (6) entirely
(7) relative (8) perform (9) soar
WORD HUNT (1) cocoa (2) o'clock (3) roar
(4) scholar (5) force (6) safety (7) Surely
(8) skull (9) utmost (10) period (11) ninety

9 WORD PUZZLE (1) weapon (2) weariness
(3) naturally (4) generally (5) sleepiness
(6) furnace (7) business (8) surface (9) soup
EXPLANATIONS (1) wasp (2) general
(3) threat
WORD PUZZLE **Clues down** (1) forehead
(2) heaviness (4) annual (5) ounce
Clues across (3) embrace (6) cough
(7) practical

10 WORD MAKER (1) region (2) permission
(3) union (4) statement (5) improvement
(6) abundant (7) elephant
WORD PUZZLE **Clues down** (1) appeal
(2) million (3) sentence (4) geography (5) biscuit
(6) colony (7) juice (8) orphan
Clues across (1) admission (5) bicycle
(9) elephant

11 WORD PUZZLE **Clues down** (2) governor
(3) conductor (4) pause (5) subtract (7) autumn
(9) depth (10) fault
Clues across (1) August (6) laundry (8) submit
(12) author (13) length
EXPLANATIONS (1) priest (2) scent (3) fierce
(4) shriek

12 WORD MATCH (1) avoid (2) moisture
(3) abrupt (4) curious (5) anxious (6) arouse
(7) trousers (8) surrender (9) material
(10) various
WORD PUZZLE **Clues down** (1) worthy
(2) surround (3) special
Clues across (4) glorious (5) rejoice (6) fury
(7) palm
WORD HUNT (1) wither (2) choice (3) lungs

13 MINI PUZZLE (1) success (2) zone (3) loose
(4) deny (5) soldier (6) pearl (7) acre
WORD HUNT (1) soldier (2) New Zealand
(3) foolish (4) debt (5) zone (6) Crooked
(7) machine (8) woollen (9) succeed (10) doubt
(11) success (12) prayer (13) marriage
(14) successful (15) carriage

14 WORD MATCH (1) festival (2) determine
(3) personal (4) portion (5) criminal
(6) introduction (7) protection (8) proportion
(9) determination (10) production (11) liberal
(12) declare

ANSWERS

WORD PUZZLE **Clues down** (2) decline
(3) major (4) traitor (6) minor
Clues across (1) fund (5) decrease (6) majority
(7) removal

15 WORD HUNT (1) announce (2) income
(3) engineer (4) accurate (5) accustom
(6) addition (7) accuse (8) pioneer (9) skeleton
(10) career (11) injury (12) ghost
EXPLANATIONS (1) injure (2) cushion
(3) instruct (4) insert (5) insult
WORD MAKER (1) keen (2) Greece (3) greet
(4) pioneer (5) accurate (6) accuse (7) accustom

16 WORD HUNT (1) consent (2) preferred
(3) wireless (4) contract (5) enterprise
(6) convince (7) wholesome (8) poison
(9) contempt (10) grateful (11) therefore
(12) coil
EXPLANATIONS (1) conclude (2) control
(3) Prefer (4) conferred (5) wisdom (6) condemn
MINI PUZZLE (1) coil (2) lodging
(3) conclude (4) grudge (5) prefer (6) contempt
(7) control

17 WORD MATCH (1) delicate (2) fragment
(3) discourage (4) hesitate (5) experiment
(6) navigate (7) remedy (8) confident
WORD HUNT (1) continent
(2) student, certificate (3) average, baggage
(4) debate (5) garage (6) candidate
(7) cement (8) confident, camera (9) regiment
(10) stage

18 WORD HUNT (1) nervous, ridiculous, couch
(2) prosperous, tremendous (3) jealous
(4) Agriculture, lecture
(5) temperate, temperature (6) puncture
(7) prominent, treatment
(8) instrument, ornament (9) volcano, recently
WORD HUNT (1) treaty, treatment
(2) recent, recently (3) route (4) tremendous
WORD MATCH (1) prosperous (2) tremendous
(3) ridiculous

19 WORD HUNT (1) appearance (2) witch
(3) responsible (4) wretched (5) rifle (6) wrinkle
(7) ignorant (8) attendance
WORD PUZZLE (1) deceit (2) ignorance
(3) visible (4) deceitful (5) invisible (6) perceive
(7) sensible (8) ceiling (9) earthquake
WORD MAKER (1) hatred (2) sacred
(3) abundance (4) appearance (5) ignorance

20 WORD HUNT (1) principal, fatal
(2) Cathedral, national (3) continually, sensation
(4) ventilation, consideration
(5) intention, punctual (6) combination, physical
(7) telescope (8) telegraph (9) telegram
(10) photograph

MINI PUZZLE (1) conversation (2) phrase
(3) choir (4) section (5) principal

21 WORD PUZZLE **Clues down** (1) vacant
(2) athletic (4) remembrance (6) domestic
(9) insurance (10) seize (11) pilot (13) tyrant
(15) ideal
Clues across (3) fragrant (5) elegant (7) tropics
(8) item (12) circumstance (13) tenant
(14) majestic (16) assistance (17) extravagant

22 WORD HUNT (1) description, probably,
reliable (2) probable, muscle, reasonable
(3) miserable, destruction (4) convey, capable
(5) cleanliness (6) witness (7) despise
(8) concern, convert (9) energy (10) liable
WORD HUNT (1) despair (2) trifle (3) angle,
cleanliness

23 WORD HUNT (1) genuine (2) crime (3) circus
(4) objection (5) instruction (6) satisfaction
(7) ambition (8) reception (9) reference
(10) residence (11) presence (12) influence
(13) pretence (14) absence (15) definite
(16) evidence (17) famine (18) medicine

24 WORD MATCH (1) destitute (2) salute
(3) distribute (4) deposit (5) volume (6) execute
(7) develop (8) gratitude (9) acquaintance
(10) acquire (11) detail (12) burden
WORD HUNT (1) acquaint (2) abbey
(3) swollen (4) strength (5) strengthen
(6) luxury (7) acquainted (8) disappoint
WORD HUNT (1) deposit (2) distribute
(3) burden (4) execute

25 WORD HUNT (1) application, response
(2) opposite, approve (3) release, remainder
(4) traveller, reveal (5) offend (6) pattern, design
(7) college, Challenge (8) messenger
WORD MAKER (1) oppress (2) oppose
(3) opposition (4) resign (5) reserve (6) reveal
WORD HUNT (1) challenge (2) release
(3) remainder (4) offend

26 WORD HUNT (1) independent
(2) occasionally, mischievous, innocent
(3) impudent, occasion
(4) numerous, advertisement, advertise
(5) marvellous, excitement
MINI PUZZLE (1) practise (2) shield
(3) advertise (4) pierce (5) yield
WORD MAKER (1) vapour (2) rumour
(3) flavour (4) enormous (5) generous

27 WORD PUZZLE (1) invalid (2) impatient
(3) opinion (4) impatience (5) operation
(6) invention (7) digestion (8) pronounce
(9) proclaim (10) civilised

46 ESSENTIAL SPELLING

WORD HUNT (1) frigid (2) patient, digest (3) proceed (4) provoke, civil (5) poultry (6) mould
WORD MAKER (1) provoke (2) proceed (3) proclaim

28 WORD HUNT (1) pension (2) recollect (3) provision, division (4) dissolve (5) recommend (6) wholly (7) shipping (8) complaint (9) solve, dissolve (10) annoyed (11) humour, endeavour
WORD HUNT (1) Comrade (2) decision (3) tobacco (4) commence (5) conclusion (6) annoy (7) commit
WORD MATCH (1) conclusion (2) wholly (3) comrade

29 OPPOSITES (1) arrival (2) original (3) variety (4) poverty (5) moral
WORD PUZZLE **Clues down** (2) dainty (3) programme (4) origin (5) translate (6) sacrifice (9) society (10) quench
Clues across (1) sandwich (7) transform (8) crystal (11) quaint (12) mutiny (13) character

30 WORD MATCH (1) obstinate (2) interfere (3) pursue (4) honourable (5) peaceable (6) scheme (7) suitable (8) create
WORD HUNT (1) Chinese (2) emigrate (3) extremely (4) manageable (5) emigrant (6) extreme
MINI PUZZLE (1) source (2) mourn (3) supreme (4) creation (5) suitable (6) scheme

31 WORD MATCH (1) incident (2) judgement (3) offence (4) quarry (5) apparent (6) maintain
WORD HUNT (1) intelligence (2) squirrel (3) intelligent (4) essay (5) difference (6) type (7) typewriter (8) portrait (9) magnificent (10) compliment
MINI PUZZLE **Clues down** (1) sustain (2) torrent (3) thorough
Clues across (4) parliament

32 WORD MATCH (1) irregular (2) similar (3) extraordinary (4) vulgar (5) salary (6) attach (7) circular (8) convenient (9) boundary (10) experience (11) existence (12) consequence
WORD HUNT (1) missionary (2) tributary (3) boundary (4) convenience (5) existence (6) obedience (7) obedient
MINI PUZZLE (1) psalm (2) military (3) fowl

33 WORD HUNT (1) disguise (2) league (3) spectacle (4) violent (5) cease (6) obstacle (7) breathe
WORD PUZZLE (1) plague (2) descend (3) scissors (4) permanent (5) awkward (6) intimate (7) science (8) conceal (9) fatigue

EXPLANATIONS (1) miracle (2) ascend (3) rear (4) disguise or conceal

34 BOYS' NAMES (1) social, artificial, especially (2) stomach (3) musician
WORD MATCH (1) precious (2) gracious (3) suspicious (4) opportunity (5) delicious (6) ability (7) artificial (8) charity (9) possibility (10) responsibility
WORD HUNT (1) wreath (2) triumph (3) Especially (4) curiosity (5) brooch (6) precious (7) suspicion (8) musician

35 VOWEL SPOTTER (1) religious (2) victorious (3) behaviour (4) expedition (5) industrious
WORD MATCH (1) serious (2) exclude (3) excursion (4) necessity (5) behaviour (6) previous (7) furious (8) necessary (9) scarlet (10) religious
EXPLANATIONS (1) religion (2) prophet (3) explode (4) exception (5) exceed (6) solemn (7) saviour
WORD MAKER (1) exceed (2) exception (3) explode (4) expedition

36 WORD HUNT (1) system (2) delivery (3) establish (4) electricity (5) sympathy (6) distinguish (7) sovereign (8) atmosphere (9) discovery (10) leisure (11) persevere
WORD PUZZLE (1) settler (2) extinguish (3) mystery (4) abolish (5) urgent (6) urge (7) interrupt (8) sovereign (9) persuade (10) diminish
WORD MAKER (1) extinguish (2) diminish (3) establish (4) distinguish

37 WORD HUNT (1) attraction (2) ruffle (3) pressure (4) scribble (5) accompany (6) twilight (7) affectionate (8) assure (9) affair (10) rubbish (11) appoint (12) hoarse, coarse
MINI PUZZLE **Clues down** (1) accomplish (4) traffic
Clues across (2) affectionate (3) summit (5) coarse
WORD MAKER (1) assume (2) assemble (3) assent (4) assure

38 WORD HUNT (1) Nursery, occupation (2) recognize, horizon (3) collection, congregation (4) champion, preparation (5) realize, novel, satisfy (6) prospect, separation (7) organization, attractive (8) cancel (9) organize, machinery (10) jewellery, marvel
WORD HUNT (1) organization (2) congregation

39 WORD HUNT (1) indulging, passion, chocolate (2) soothe (3) possession, immediately, celebration (4) illustrate (5) imitate, imitation

ANSWERS

WORD MATCH (1) correspond (2) interior
(3) inferior (4) discussion (5) exterior
(6) superior (7) senior (8) indulge
WORD HUNT (1) impression (2) immediately
(3) correspond (4) celebrate, illustrate
WORD MAKER (1) senior (2) interior
(3) inferior

40 WORD PUZZLE **Clues down** (2) applaud
(5) suburb (7) audience (8) nourish (9) exhibit
(10) benefited (11) register (12) launch
(13) exhaust (15) benefit (16) council
Clues across (1) partner (3) privilege
(4) cautious (6) profited (7) album
(14) authority (17) exhibition (18) clause
(19) breadth

41 WORD HUNT (1) instructor (2) postpone
(3) radiator (4) liquid (5) acknowledge
(6) nonsense
WORD PUZZLE Clues down (1) haughty
(2) curve (3) suspense (4) envelope (5) conquer

Clues across (3) slaughter (6) intense
(7) condense (8) disease (9) conqueror
WORD HUNT (1) dense (2) immense
(3) knowledge

42 WORD MATCH (1) association (2) column
(3) brilliant (4) positive (5) peculiar (6) associate
(7) appreciation (8) sufficient (9) ancient
(10) scorch
WORD HUNT (1) ancient (2) stubborn
(3) survive (4) horror (5) Christian (6) official
WORD HUNT (1) Spaniard (2) familiar
(3) hymn (4) Spanish

43 WORD HUNT **A** (1) siege (2) Egypt
(3) pitiful (4) Courageous (5) succession
(6) welfare **B** (1) procession (2) ceremony
(3) courteous (4) fulfil (5) museum (6) theatre
C (1) unusual (2) salmon (3) skilful
(4) absurd (5) virtue (6) suggest
WORD MAKER (1) issue (2) rescue (3) virtue
(4) skilful (5) pitiful (6) courteous
(7) courageous